Planter & Tree-Guard   Bike Rack   Public Recreational Facilities

PUBLIC LANDSCAPE
STREET FURNITURE I

· Pavement   Seat   Rest Pavilion   Drinking Fountain

地景设施 上

高迪国际 HI-DESIGN PUBLISHING   编

王丹 李小童 刘宪瑶 陈曦   译

广西师范大学出版社
·桂林·

图书在版编目(CIP)数据

地景设施:汉、英／高迪国际 HI-DESIGN PUBLISHING
编;王丹,李小童等 译. —桂林:广西师范大学出版社,
2014.8
ISBN 978 - 7 - 5495 - 5431 - 7

Ⅰ. ①地… Ⅱ. ①高… ②王… ③李… Ⅲ. ①城市景
观-景观设计-作品集-世界-现代 Ⅳ. ①TU - 856

中国版本图书馆 CIP 数据核字(2014)第 100795 号

出 品 人:刘广汉
责任编辑:肖　莉
装帧设计:高迪国际 HI-DESIGN PUBLISHING
广西师范大学出版社出版发行

(广西桂林市中华路 22 号　　邮政编码:541001)
(网址:http://www.bbtpress.com)

出版人:何林夏
全国新华书店经销
销售热线:021 - 31260822 - 882/883
上海锦良印刷厂印刷
(上海市普陀区真南路 2548 号 6 号楼　邮政编码:200331)
开本:646mm×960mm　　1/8
印张:64　　　　字数:45 千字
2014 年 8 月第 1 版　　2014 年 8 月第 1 次印刷
定价:536.00 元(上、下册)

PREFACE 序言

Public Realm

What happens when we cross the threshold of our house and immerse ourselves in the public realm? Can we extend that domestic feeling into public space allowing the possibility of intimacy, of encounter and integration?

There are many appreciations of what public space is. Spanish architect Juli Capella defines public space as a great "U" being the "U" the place defined by its roads and façades in which public life goes by.

As a South American architect and designer I prefer referring to it as the urban "O" adding to roads and façades the concept of roof which, materialised by the tree canopies or through the presence of the sky shapes this great container to inhabit, which is public space.

Public space is strongly bound to a city's cultural identity, to its history, to its traditions: the urban elements and communitarian support systems play an important role in the definition of that identity, qualifying it, whichever the scale.

Through repetition and interaction the urban elements and supports create a wide variety of services that perform as a background for urban life. From micro architecture to pavement patterns to rest elements and urban fittings they conform a sensible fabric of facilities and establish a physical communication with all the community. They constitute a sensible skin where the community inhabits.

This skin has the mark of history as an active factor, footprints of materials, technology and use, all define the mark of past and present and as such superimpose in permanently evolving spatial and temporal contexts.

Mankind changes socially, and so do rites and ways of celebrating encounters, communion or even solitude. As interpreters of the new "urban tribes" and their social rites, we must adapt to their necessities and delve into the particularities of their behaviour.

The enjoyment of an exterior urban space as a counter balance to the intensity of urban life allows us to reformulate the well-known concept of each type of urban elements. We are therefore inclined to re-imagine public elements as a flexible and neutral open support system.

Thus, the elements and objects should lose their enunciation and become more oriented towards the use and quality of the site and its corresponding appropriation, instead of particular elements with recognisable illustrative names and a pre-established images or typologies. The result could be the reinterpretation or tangible representation of a verb; to reunite, to rest, to take refuge, to relax.

Social rites define choreographies, relationships between space and time, and it is the job of us designers to attempt to interpret them.

Designing in the city requires from us a great challenge in interpreting the required scale of intervention. The city's support system can be conceived either from a generic or a particular perspective.

Addressing the city implies on the one side a most global approach which involves active supports oriented towards the construction of urban mobility and connectivity, and on the other side, a more particular approach, where supports are of a passive nature and interpret the social structure in terms of community and neighbourhood scales.

The urban intervention, far from being a rational and abstract operation, should be based on an anthropological approach founded upon the idea of inhabiting landscape, either in an individual or a communitarian manner.

Exploring the basic conditions of inhabiting a natural or urban landscape as in a mirror game, landscape collides with us inhabiting our own body.

公共领域

当我们踏出家门，进入公共领域时，会发生什么呢？

我们能否将家的轻松氛围延伸到公共空间、与人相遇、相知，相处融洽？

人们对公共空间的理解多种多样。

西班牙建筑师胡利·嘉佩乐将公共空间定义为一个大大的 "U"，这个 "U" 由道路和建筑外观共同构筑而成，人们在这个 "U" 中生活作息。

作为一名来自南美洲的建筑师和设计师，我却更愿意在道路和建筑外观上增添顶盖的概念，将公共空间定义为 "O"。顶盖或是树冠或是天空，围合成巨大的居住空间，这就是公共空间。

公共空间与城市的文化特征、城市的历史与传统关系密切：无论在哪一个层次，城市设施和社区支撑体系对城市文化特征的勾画都起到重要作用，并对其进行限定。

城市设施和社区支撑系统通过重复和相互作用建立起一系列丰富多样的服务，支撑着城市生活。无论是微观建筑、人行道图案，还是休息场所，抑或是城市装置，均构成实用的设施构架，且与整个社区和谐共存。

这些赋予社区一个很明显的外在表现。

社区的外表体现出历史这一活跃因素的痕迹、材料的印记，也体现出技术与应用。所有的一切都是过去与现在的映像，因此也就随着时空的不断发展变化而变化。

人类的社交方式在不断变化着，礼节仪式、庆祝邂逅的方式、沟通交流的方式，甚至人们独处的方式都在发生改变。我们作为这新型 "都市部落" 及其社交仪式的阐释者，必须作调整以适应其规律，研究其行为上的细节特征。

我们既可以在城市外部空间找到乐趣，也可以体验到紧张的城市生活，两者可以互相制衡，这就允许我们重新思考各种已经约定俗成的城市设施理念。于是，我们更喜欢进行重新设想，把公共设施视为一个既灵活又中立开放的支撑体系。

因此，在设计时，应该摒弃城市元素和设计客体的突兀之处，更关注场所的使用、品质以及其与周围的和谐度，而不是关注那些众所周知的学术名词或知名度高的形象或者是某种经典类型。结果有可能是设计师要重新诠释或对某一动作进行再现：设计是使人们有地方团聚、休息、寻求庇护或放松。

社区礼仪明确了时空之间的艺术和关系，而我们设计者的工作就是试着去解释时空之间的关系。

城市设计要求我们在阐述必要的介入规模时面临巨大挑战。城市的支撑系统可以从一般与特殊两个方面来考虑。

装点城市，意味着一方面要采用整体法，要以积极支持的方式实现城市建设的灵活性和连贯性；另一方面，要采用具体法，即具体问题具体分析，以社区和街道的规模阐述社会结构。

城市建设远非理性的、抽象的操作，而应该以人类学的方法为基础，建立在居住景观理念之上，无论对个体还是对群体而言都是如此。

无论居住在自然环境或城市环境，探索生活的基本条件，就像在玩镜子游戏一样，环境与我们合二为一，和谐共生。

戴安娜·卡维萨

Diana Cabeza

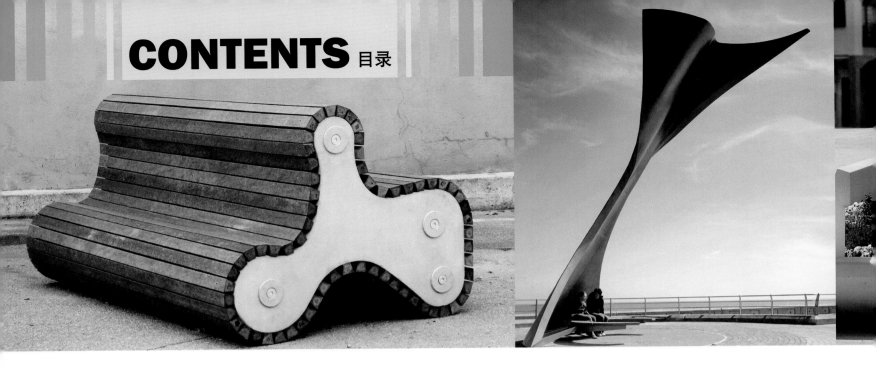

CONTENTS 目录

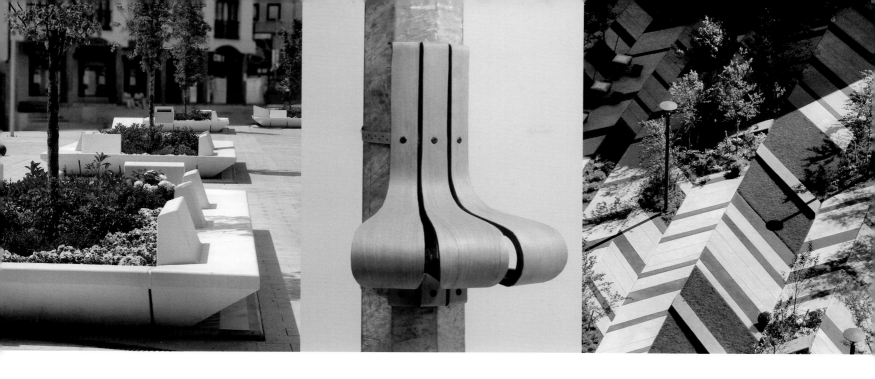

PLANTER & TREE-GUARD

花坛护树设施

Wrap Planter

卷型花架

欧几里得长廊项目斥资数百万美元，而 Mark A. Reigelman 二世应邀设计其中一部分装饰性花架。花架的形状使得街景充满活力，而该设计使人想起从街边购买的裹着花束的包装纸。"我想赠与每一个克利夫兰的行人一束新鲜的花，" Reigelman 说。

这些花架的设计考虑了季节性种植，增强了克利夫兰市中心欧几里得大街的视觉效果。通过市中心正在进行的美化项目，LAND 工作室主要负责这些花架的设计和安装。

目前在克利夫兰市中心有 200 多个卷型花架。这些花架由钢筋混凝土制成，约 1.2 米高，680 公斤重。

Mark A. Reigelman II was commissioned to design decorative planters as part of the multi-million dollar Euclid Corridor Project. The shape of the planters animates the ridged streetscape and the design suggests the paper wrapping that surrounds a bouquet of flowers purchased from a street vendor. "I wanted to offer every Cleveland pedestrian a fresh bouquet of flowers," says Reigelman.

The planters are designed for seasonal plantings that enhance the visual character of Euclid Avenue in downtown Cleveland. Through its ongoing downtown beautification programme, LAND studio has taken the lead on the design and installation of the plantings.

There are over 200 wrap planters currently in Downtown Cleveland. The planters are made of steel reinforced concrete and are approximately 4' in height and weight almost 1,500 lbs.

DESIGN COMPANY	DESIGN TEAM	PROJECT ARCHITECT	LOCATION
Tonkin Liu	Mike Tonkin, Anna Liu, Robert Urbanek-Zeller	Robert Urbanek-Zeller	London, UK

Promenade of Light

阳光长廊

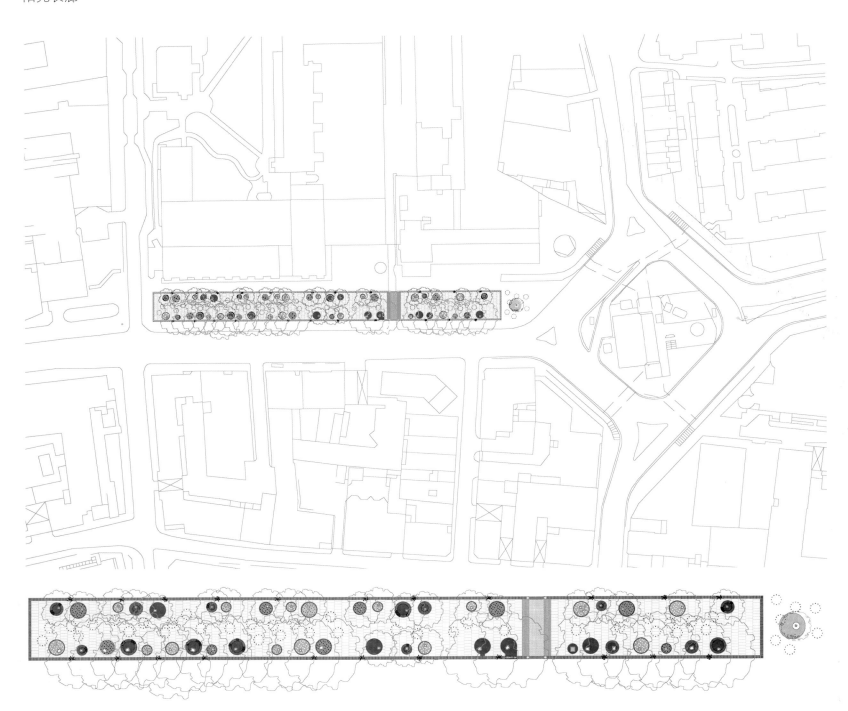

The brief was for the improvement of a grassed area in front of the shops, with strategic proposals for the areas surrounding the Old Street Roundabout.

Promenade of Light

Walking alone, ideas unfold, walking with others, relationships grow. A promenade is the very celebration of walking, and of the community. The existing twenty-one mature Plane trees suggested a promenade that could be reinforced by adding more trees. Eighteen new trees were added, and a raised stone promenade laid between the two rows of trees. The light stone is a canvas for light and shadows, during both day and night.

Celebrating Difference

Around the trees, stone rings protect the trees and mark them as special.

Each ring is different, some big, some small, and some in-between. Rings for people are benches and tables. Rings for plants are filled with brightly coloured flowers. Clusters of rings bring people together and create urban rooms under the shelter of the trees.

Space formed with Light

The proposal includes several ways of lighting, to back-light the leaves, to cast shadows of people and trees, to project circles of dappled light on the paving, to spot-light the flowers and tree trunks. Twenty-three lamp poles, each with six to eight lamps, light the space from high level. From 5-degree to 100-degree beam angles, and from yellow safety light to bright white light, each lamp was positioned individually, at a specific element on the promenade. Sun-tracking timed switch has been programmed to vary between weekends and weekdays, summer and winter.

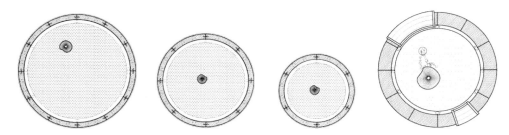

large planter   medium planter   small planter   large bench

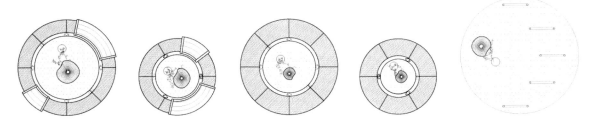

medium bench   small bench   medium table   small table   cycle stands

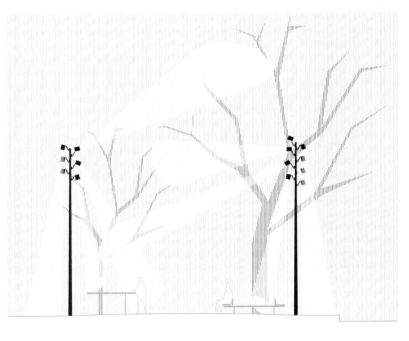

阳光长廊

独行者，思路通达；结伴者，谈笑风生。一条长廊既方便了步行，又推动了社区交流。现有的 21 棵高大梧桐树构成了长廊风貌，今后会种植更多的树木加强这种风貌。18 棵新树将被种植，一条突起的通道铺在两排树木之间。白色的砖面成为光与影的幕布，在白天和黑夜上演一幕幕场景。

特色圈状桌椅

石圈围绕着树木，既是对树木的保护也是每棵树木的特殊标识。环圈各有不同，有的大，有的小，还有些面积适中。圈环对于人们既可以当长椅也可作桌台。为灌丛建造的圈环中往往生长着色泽鲜艳的花朵。环环相扣的圆圈将人们联系在了一起，在树荫的遮蔽下创造出都市休闲空间。

光影空间

方案包括多种光照模式，灯光从树木外层照射进长廊，在石板路上映出行人和树木枝叶的阴影，并在路面投下一圈圈斑驳的明亮，在树干和花丛上也有星星亮点。23 盏路灯，每盏有 6~8 个灯泡，在高处照射长廊。从 5 度到 100 度的光束角，从黄色的安全灯到白色的照明灯，每个灯都被摆放在不同位置，成为长廊的特殊要素。

太阳跟踪定时开关已被编程在周末和平日、夏季和冬季之间变化。

DESIGNER
Manuel Ruisánchez

MANUFACTURER
Escofet

CHARACTERISTIC
Reinforced cast stone, Standard colour chart, Acid-etched and waterproofed. Free-standing

Lena

花槽 Lena

Ruisanchez Architects have developed LENA, a large-volume planter with an outer diameter of 230 cm, based on the geometry of a slanted cone, sectioned by two non-parallel planes. The surface of this cast stone product with a mild acid etch features a deep grooved relief, highlighted by a play of light and shadow that gives it strong personality. Two positions are possible: LENA-open, with the largest diameter at the top, and LENA-closed in the reverse position. In both cases, the rear part is open. LENA can be installed directly on pavements, sand or natural ground. When the interior volume cannot be placed in direct contact with the natural ground, the soil volume in the planter is confined with a double layer of geotextile or a thin base layer of porous concrete to facilitate the drainage of excess water without soil loss or erosion.

PHOTOGRAPHERS
Anna Pericas, Shlomi Almagor

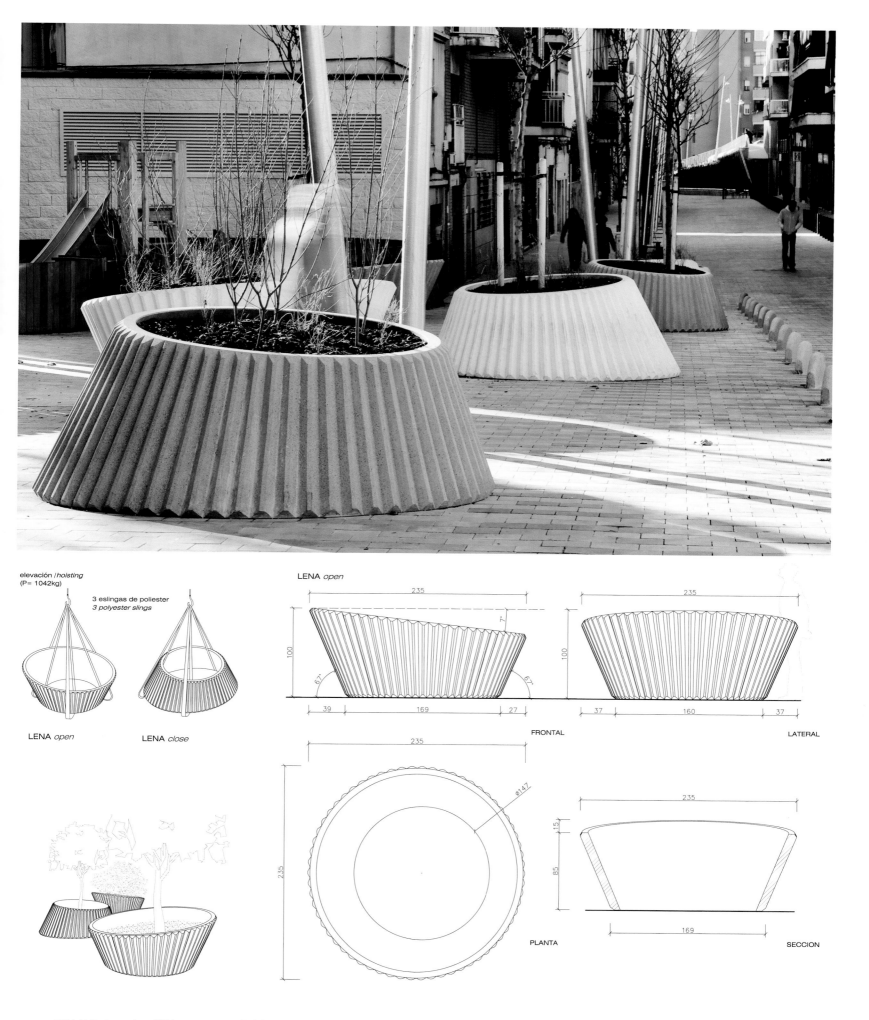

elevación / hoisting
(P= 1042kg)

3 eslingas de poliester
3 polyester slings

LENA open

LENA close

LENA open

235

100

7

67°

39   169   27

FRONTAL

235

100

67°

37   160   37

LATERAL

235

Ø147

235

PLANTA

235

15

85

169

SECCION

建筑师 Ruisanchez 设计了 LENA: 一个外径为 230 厘米的大体积的花槽, 以几何斜锥形为基础, 切面为两个不平行的平面。此铸石产品表面用温和的酸液腐蚀过, 形成了一条条沟槽浮雕, 突出了明和暗的对比, 赋予了它强烈的个性。两个方向都很合理: 开口型 LENA 的最大直径在上平面, 而闭口型 LENA 则与其相反。在这两种情况下, 后一部分是开口型的。LENA 可以直接安置在人行道上、沙子上或天然地表上。当内部容积不能放置在与自然地表直接接触的地方时, 花槽里的土壤体积就局限于一个土工织物或者一个薄的基础层的多孔混凝土, 让多余的水分排出来, 而泥土不会流失也不会遭受侵蚀。

DESIGN COMPANY

FÜNDC

DESIGN TEAM

César García Guerra, Paz Martín Rodríguez, Juanjo Unceta Rivas, Marina Otero, Fatima Plaza, Liu Pei, Zuoming Wang, Julia Rodriguez Buján,Diego Ochayta, Gema Edo Viñarás, María Nieves González San Millán, Alexandra Moreno Arranz, Johan de Wachter, Patricia Mata.

Nuevo Centro Cultural - Mega-Treepots

新文化中心 (NCC)-mega- 树木盆栽

STRUCTURE

White structural concrete for mega-treepots

LOCATION

Madrid, Spain

PHOTOGRAPHER

César Gª Guerra © FÜNDC

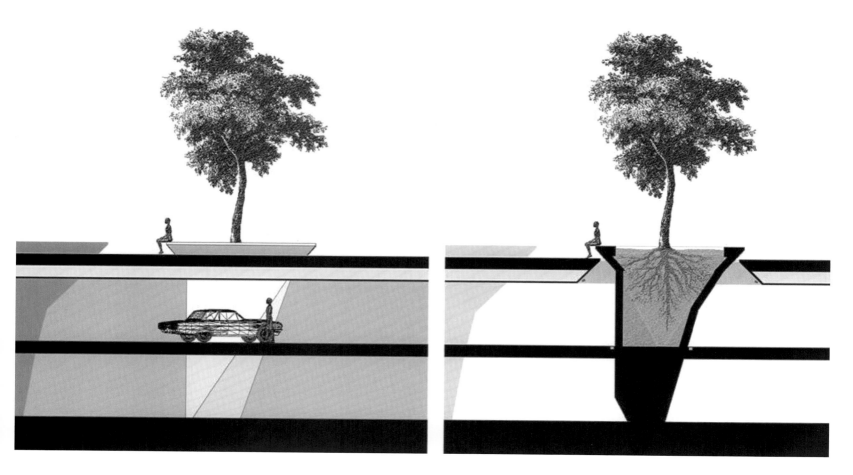

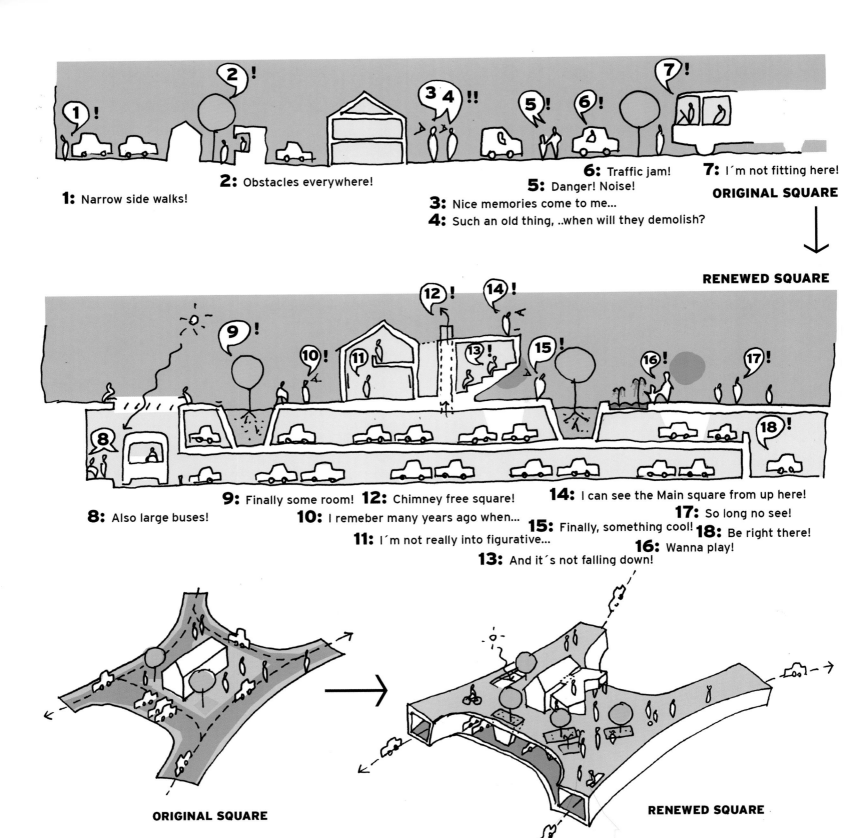

1: Narrow side walks!
2: Obstacles everywhere!
3: Nice memories come to me...
4: Such an old thing, ..when will they demolish?
5: Danger! Noise!
6: Traffic jam!
7: I´m not fitting here!

ORIGINAL SQUARE

RENEWED SQUARE

8: Also large buses!
9: Finally some room!
10: I remeber many years ago when...
11: I´m not really into figurative...
12: Chimney free square!
13: And it´s not falling down!
14: I can see the Main square from up here!
15: Finally, something cool!
16: Wanna play!
17: So long no see!
18: Be right there!

ORIGINAL SQUARE

RENEWED SQUARE

SECCION J7J8-L
1:125

O22

The pots allow for the growth of medium-large trees above an underground parking, making possible green areas where usually just hard squares are found.

树盆可以使中型树木生长在地下停车场之上，尽可能地在地上创造出绿色空间，这些绿地通常只能在广场上觅得踪迹。

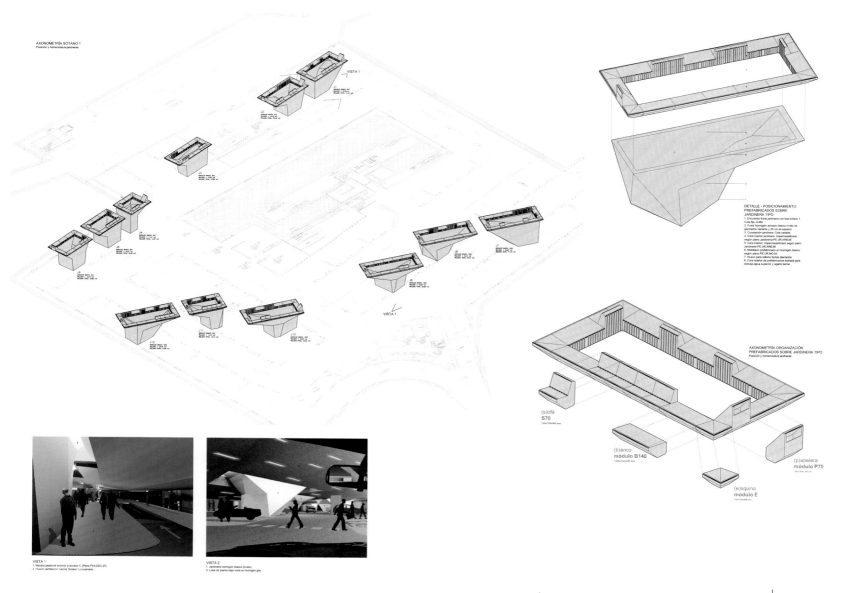

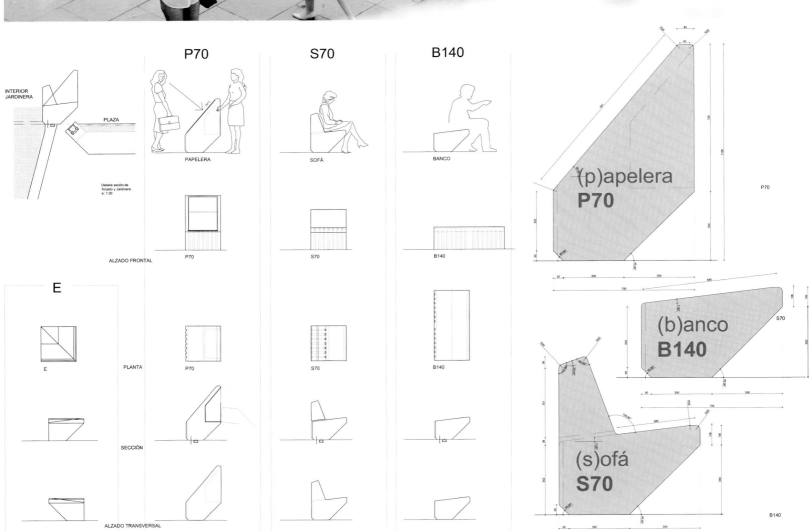

INTERIOR
JARDINERA

PLAZA

Detalle sección de
forjado y Jardinera
e: 1:20

P70

PAPELERA

ALZADO FRONTAL

P70

S70

SOFÁ

S70

B140

BANCO

B140

E

PLANTA

P70

S70

B140

SECCIÓN

ALZADO TRANSVERSAL

(p)apelera
P70

P70

(b)anco
B140

S70

(s)ofá
S70

B140

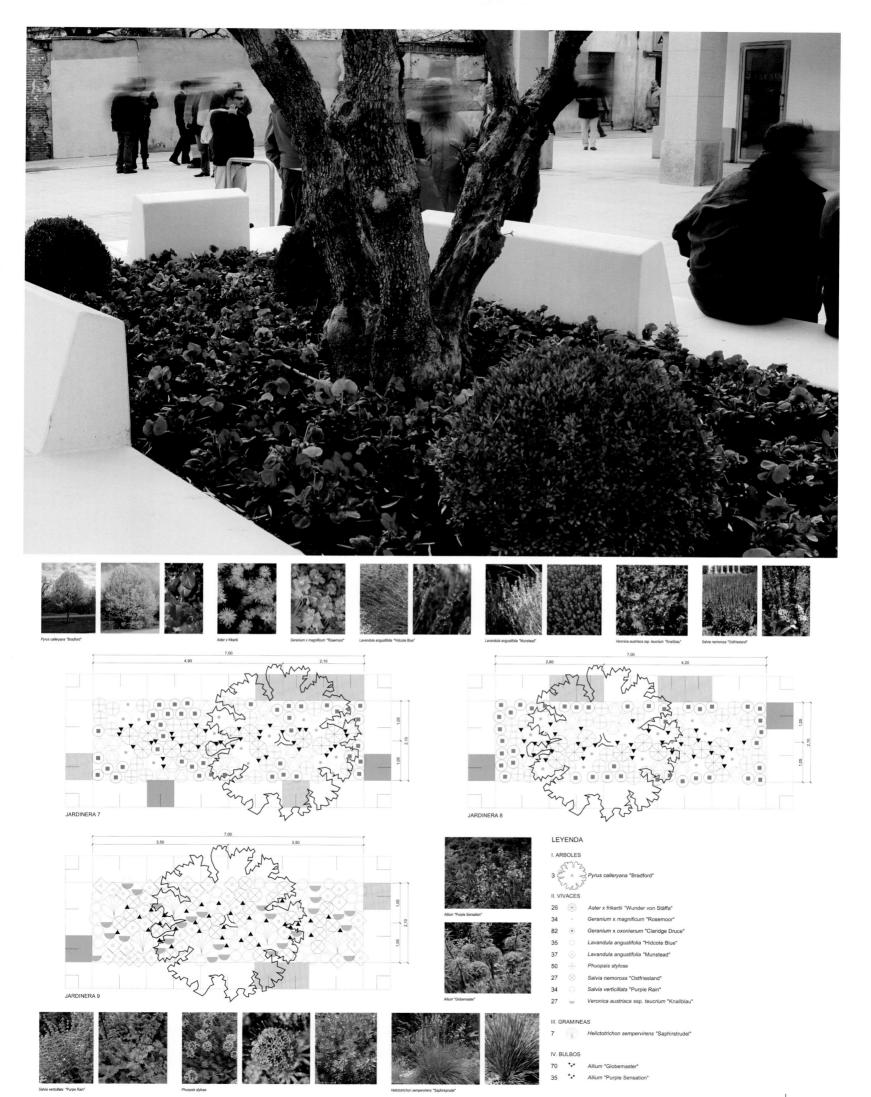

Pyrus calleryana "Bradford"   Aster x frikartii   Geranium x magnificum "Rosemoor"   Lavandula angustifolia "Hidcote Blue"   Lavandula angustifolia "Munstead"   Veronica austriaca ssp. teucrium "Knallblau"   Salvia nemorosa "Ostfriesland"

JARDINERA 7

JARDINERA 8

JARDINERA 9

Allium "Purple Sensation"

Allium "Globemaster"

LEYENDA

I. ARBOLES

3   Pyrus calleryana "Bradford"

II. VIVACES

26   Aster x frikartii "Wunder von Stäffa"
34   Geranium x magnificum "Rosemoor"
82   Geranium x oxonianum "Claridge Druce"
35   Lavandula angustifolia "Hidcote Blue"
37   Lavandula angustifolia "Munstead"
50   Phuopsis stylosa
27   Salvia nemorosa "Ostfriesland"
34   Salvia verticillata "Purple Rain"
27   Veronica austriaca ssp. teucrium "Knallblau"

III. GRAMINEAS

7   Helictotrichon sempervirens "Saphirstrudel"

IV. BULBOS

70   Allium "Globemaster"
35   Allium "Purple Sensation"

Salvia verticillata "Purple Rain"   Phuopsis stylosa   Helictotrichon sempervirens "Saphirsprudel"

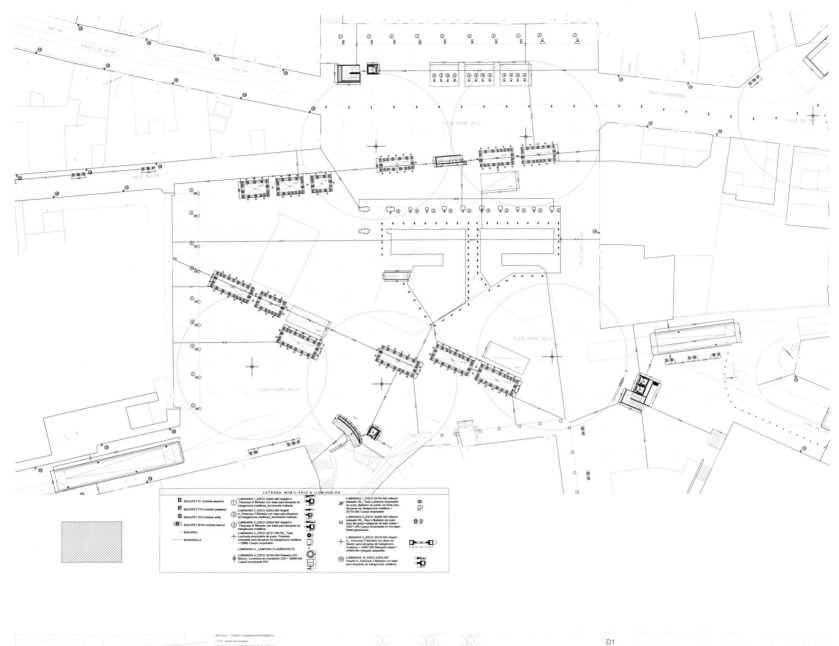

LEYENDA MOBILIARIO E ILUMINACION

E   ESCOFET E1 (módulo esquina)
P   ESCOFET P70 (módulo papelera)
S   ESCOFET S70 (módulo aula)
B   ESCOFET B140 (módulo banco)

○   BOLARDO
- - -   BARANDILLA

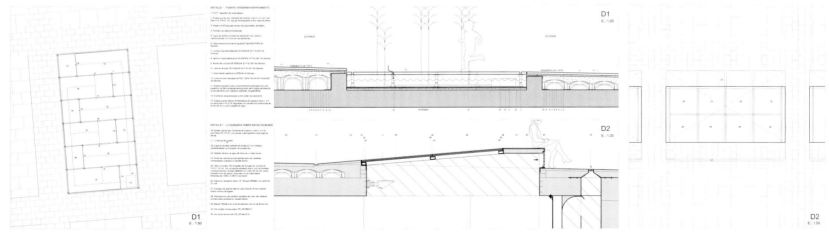

DETALLE 1 : FUENTE INTEGRADA EN PAVIMENTO   D1
                                                                                 E: 1:20

DETALLE 1 : LUCERNARIO SOBRE ESTACIONAMIENTO   D2
                                                                                 E: 1:20

D1   D2
E: 1:50   E: 1:50

DESIGNER
Ignacio Ciocchini, IDSA. Industrial Designer

MANUFACTURER
Landscapeforms / Studio 431

LOCATION
Chelsea, New York, USA

Trees and Tree-Guards

树木和树木护栏

Problem: There was a lack of vegetation in the area and the sidewalks were very harsh and unwelcoming.

Solution: Trees with garden areas were planted to soften the urban environment. Each tree includes a 10'x5' urban garden area around it with a tree-guard that protects it. The design of the tree guards relates to the structures of the nearby High-Line Park creating a sense of place for the neighbourhood. They also provide protection from pedestrians, that can't walk on them, salt, dogs, sidewalk garbage, and corrosive liquids that are sometimes deposited on sidewalks by pedestrians and other users.

• An in-depth subsurface survey was done. Sixty-eight trees of five different species were planted.

• 10'x5' planting areas allow for extra vegetation and natural irrigation for the trees.

• Belgium blocks were installed when the sidewalk grade was too steep to facilitate grade changes.

• The tree-guards that protect the tree vary in size to accommodate different sidewalk widths.

MATERIAL
Powder-coated carbon-steel plates and stainless-steel bars and tubes

DIMENSIONS
10 feet x 5 feet x 18" High

PHOTOGRAPHER
Ignacio Ciocchini

问题：该地区缺乏植被覆盖，人行道粗糙且不美观。

解决方案：在花园里种植树木给城市环境增加柔和因素。每一棵树周围都包括一个 10 寸 x5 寸的花园区域，用树木护栏来保护它。树木护栏的设计联系到附近的 High-Line 公园的建筑结构，给社区创造一种地域感。围栏保护树木免受路人破坏，行人禁止在上面行走，因为行人和其他使用者有时会将盐类食品、宠物狗、垃圾、腐蚀性液体遗留在人行道上。

- 一项深入地下的调查完成。已经栽种 68 棵 5 个不同品种的树木。

- 10 寸 x5 寸种植地区的树木允许额外的植被和自然灌溉。

- 安装比利时砖是防止人行道太陡，便于级数的变化。

- 树木护栏用来保护树，大小可进行变化以适应人行道的不同宽度。

BIKE RACK
自行车摆放架

DESIGNER
Ignacio Ciocchini, IDSA. Industrial Designer

CLIENT
NYC Department of Transportation

MANUFACTURER
Landscapeforms / Studio 431

Q - CityRack

O 形脚踏车停放架

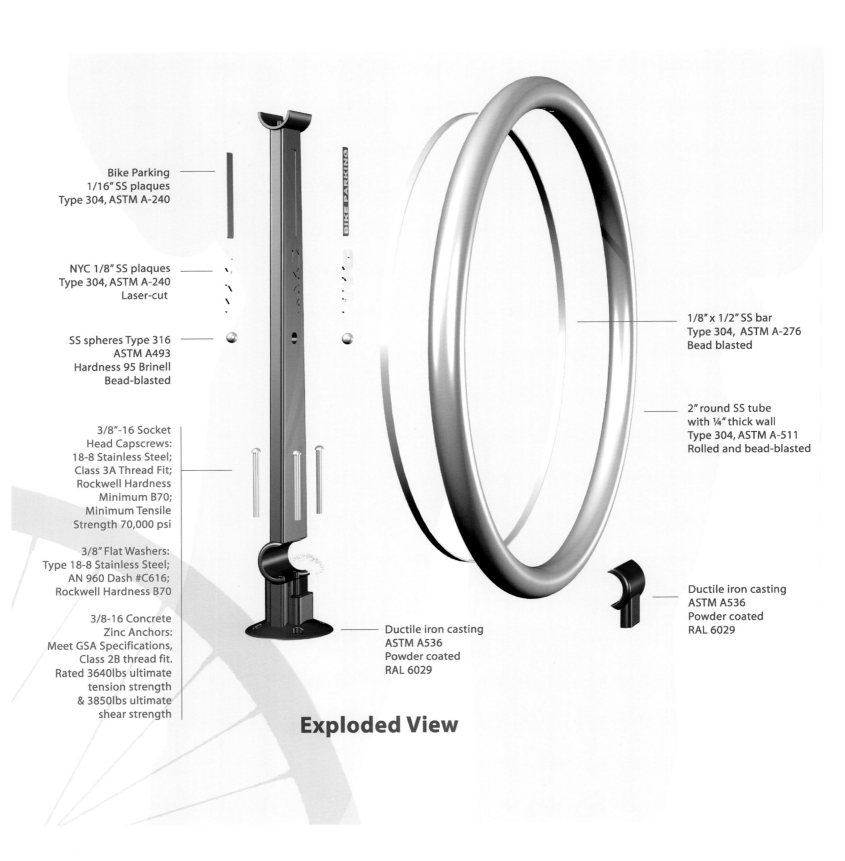

Bike Parking
1/16" SS plaques
Type 304, ASTM A-240

NYC 1/8" SS plaques
Type 304, ASTM A-240
Laser-cut

SS spheres Type 316
ASTM A493
Hardness 95 Brinell
Bead-blasted

3/8"-16 Socket
Head Capscrews:
18-8 Stainless Steel;
Class 3A Thread Fit;
Rockwell Hardness
Minimum B70;
Minimum Tensile
Strength 70,000 psi

3/8" Flat Washers:
Type 18-8 Stainless Steel;
AN 960 Dash #C616;
Rockwell Hardness B70

3/8-16 Concrete
Zinc Anchors:
Meet GSA Specifications,
Class 2B thread fit.
Rated 3640lbs ultimate
tension strength
& 3850lbs ultimate
shear strength

1/8" x 1/2" SS bar
Type 304, ASTM A-276
Bead blasted

2" round SS tube
with ¼" thick wall
Type 304, ASTM A-511
Rolled and bead-blasted

Ductile iron casting
ASTM A536
Powder coated
RAL 6029

Ductile iron casting
ASTM A536
Powder coated
RAL 6029

Exploded View

GENERAL DIMENSIONS
Height: 37 inches, Diameter: 33 inches

LOCATION
New York, NY

PHOTOGRAPHER
Marco Castro

The O Bikerack was designed and developed for the Cityracks International Design Competition that the Department of Transportation of the City of New York held in 2008. The O Bikerack was awarded 3rd place among hundreds of entries from all over the world. Two prototypes of this design are still in use in New York City today and other cities in the U.S. are considering adopting it.

The O Bikerack provides several points of contact between the bikerack and the bike frame to facilitate locking up to two bicycles using standard off-the-shelf locks and chains. The bike's frame and wheels can be connected to the bikerack ring at different points and heights, providing plenty of flexibility for the user. This feature allows the biker to make sure nobody is going to steel parts of the bike while it is locked to the bikerack.

O 形脚踏车停放架是为 2008 年纽约交通局举办的城市脚踏车停放架设计大赛所设计建造的。该设计从全球数百万的作品中脱颖而出，赢得第三名的成绩。在纽约，参赛的模型至今仍然被使用，美国的其他城市也正在考虑引入 O 形停放架的计划。

停放架提供了很多停靠点，可以轻而易举地用普通锁链将两台自行车固定住。在不同的方位和高度上，脚踏车的车架或车轮都能与停车架锁在一起，因此给使用者提供了很大的灵活性。这也使自行车的主人确信只要他们把车子锁在停车架上就没人可以偷走车子的任何一部分。

DESIGNER	DESIGN COMPANY	MANUFACTURER	MATERIAL
Jason Flannery	Forms+Surfaces	Forms+Surfaces	Bodies of Stainless Steel Tubing with heads of Cast Stainless Steel

Bike Garden Bike Racks

花样式自行车架

Landscape Architects Wood+Partners worked on the Atlanta BeltLine and specified 120 of Forms+Surfaces' Bike Garden Bike Racks, which have been installed in groupings throughout the park. Unlike anything else on the market, the Bike Garden is comprised of individual "stems" in six heights that allow you to create project-specific layouts while providing multiple locking points and secure bike parking.

The Bike Garden Bike Racks are getting plenty of use at this new park, and add a modern, sculptural aesthetic while serving a functional purpose.

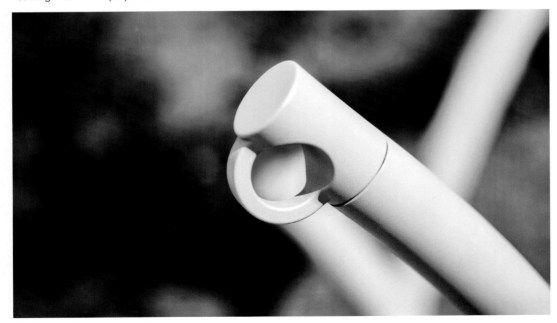

　　景观设计师Wood+Partners效力于 Atlanta Belt Line 公司，并明确了 120 种形状和表层的花样式自行车架，这些车架被成群地安置在公园里。不同于市面上的其他产品，花样式车架是由6个高度的单独的"茎条"组成的，提供多个锁点和安全停车的同时，你也可以任意摆置自行车。

　　花样式自行车架在这个新公园里得到大量的使用，并增添了现代雕塑般的美感，也使其具有功能性作用。

LOCATION
Atlanta, Georgia USA

PHOTOGRAPHER
Forms+Surfaces

DESIGNER
STORE MUU design studio

PHOTOGRAPHER
Daisuke Ito

Pit in

Pit in 脚踏车停靠桌

"PIT IN" is a table for a bicycle whose saddle functions as a chair.

Sitting on the saddle of bicycle, for example, you can take a coffee break, check e-mails by lap-top, and so on.

Enjoy a time riding a bicycle!

Even if you don't ride a bicycle, you don't mind.

"PIT IN" will be an open table for standing people.

Once "PIT IN" appears, that place will show the characteristics as a public space.

This table will open up a new lifestyle of bicycle.

Prototype 1: Steel
    Material Plywood t18
    Dimensions L115xD100xH115cm
Prototype 2: Wood
    Material Plywood t18
    Dimensions L71xD90xH115cm

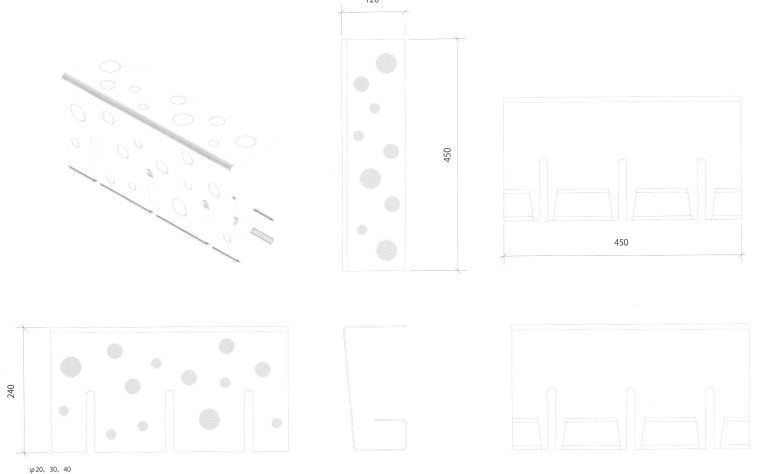

120

450

450

240

φ20、30、40

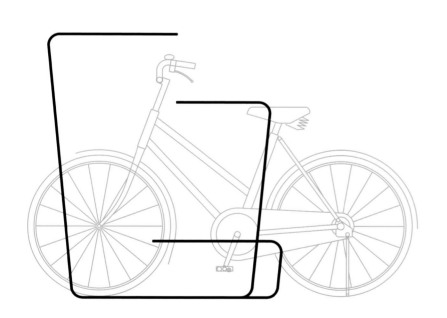

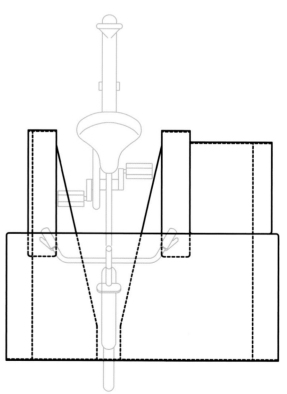

"Pit in" 自行车停靠桌是专为自行车设计的桌子，自行车车座拥有跟椅子一样的功能。

坐在车座上你可以用停靠桌做很多事情，例如，享受咖啡时光，用笔记本电脑查邮件，等等。

"在自行车座上享受一段美好时光！"

如果你没有骑自行车，一样可以。

"Pit in" 脚踏车停靠桌可以作为公用的桌子，提供给站着的人们。

一旦 "Pit in" 脚踏车停靠桌出现，就表明了该地点是典型的公共区域。

停靠桌将会开创一种关于脚踏车的新的生活风尚。

DESIGNER
David Karásek, Radek Hegmon

DESIGN COMPANY
mmcité

Meandre

曲线自行车架

This double-sided stand with an entirely new unique design in a distinct and naturally elegant shape is unprecedentedly bike-friendly. Its characteristic meander made from a durable rubber-strip can softly but firmly clasp the front or back wheel of any size. The tube running through helps to lock the bike.

这个双面的站架带着全新的、独一无二的设计，它独特而自然的优雅造型是前所未有的，便于自行车的存放。它特有的曲线由橡胶条制作而成，柔软却牢固地锁住任何尺寸的前轮或者后轮。穿透车架的管子可以让自行车锁牢。

DESIGNER	DESIGN COMPANY	MATERIAL
Margus Triibmann	Keha3	Plastic-covered metal wires, metal loops

Grazz

草丛式车架

Bicycle rack consists of plastic-covered metal wires with metal loops at the ends for fastening the bicycle locker. Plastic surface cover and elastic structure protect the bicycle from scratches. The bicycle stands upright thanks to rigid metal bars, and a small-size padlock can be used besides an ordinary bicycle locker.

Grazz bicycle racks can be positioned side by side and hence create articulations, artificial barriers to the city space.

There are two options for fastening the bicycle racks: bolted or fastened with wedge anchors to the ground or cast into the concrete.

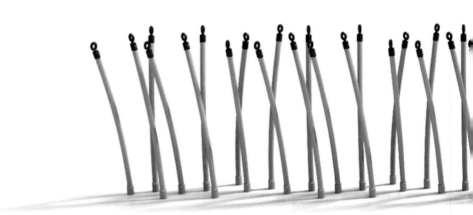

草丛式自行车架由塑料包裹的金属缆线做成，每根"草"的顶端焊接有一个金属环，用于固定自行车锁。用塑料包裹的弹性缆线是为了防止车辆的划伤。坚硬的金属底板可支撑自行车直立存放，小锁扣可以跟市面上的自行车锁相扣。

草丛式车架可以一个挨一个地排列放置，给城市空间创造了人工节点屏障。

有两个方法可以让该车架固定起来：在地面与车架底板之间用螺钉锚固，或者用混凝土浇注在底板之上。

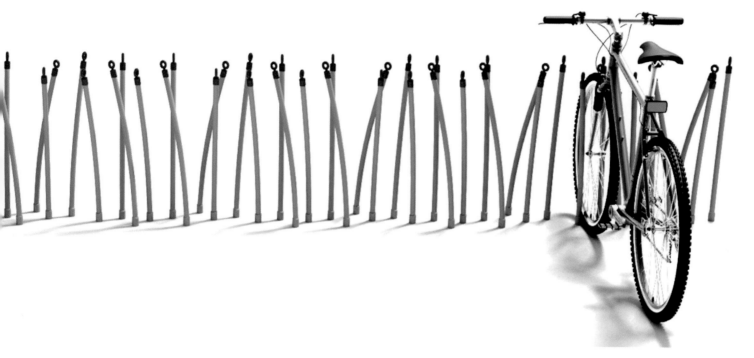

DESIGN COMPANY
Springtime

Bikedispenser

自动化自行车存储间

The Bikedispenser is a fully automated, space-saving storage and dispensary system for bicycles. It was developed to increase ease-of-accessibility to those business areas lying within three kilometres of commonly used public transportation junctions. Although the initial target-users for the Bikedispenser are indeed commuters, its ease-of-use, simplicity, adaptability and cost-effectiveness make it applicable to a multitude of recreational purposes.

Compact and Simple

The Bikedispenser is the most compact storage system for bicycles in the world, with specially designed bicycles positioned a mere 16cm apart from each other. This makes the cost for the system an amazing 70% lower than other known automated bicycle storage systems.

Security

Each user's identity is connected to an electronic ID tag (RFID) located within the bicycle. This has been shown to greatly reduce problems in terms of theft and vandalism, in contrast to the infamous 1970s Amsterdam Witte Fietsenplan (White Bicycle Plan), which was a totally anonymous system. This successful approach is currently used by OV Fiets (Dutch Rail) and Call-a-Bike (Deutsche Bahn).

Environmental Benefits

The Bikedispenser capitalises on the idea of the bicycle as the missing, final link in the commuter's mobility chain. The central idea is to create better alternatives to those imperfect solutions currently in use (such as public transportation or cars). Consumer reluctance to use bicycles to close the mobility chain is overcome by the Bikedispenser because of its ease-of-use, the availability of a good bicycle at all times and the elimination of the need to carry folding bicycles on public transportation. The Bikedispenser takes very little space, and can even be placed under the ground.

自动化自行车存储间是一个全自动的、节约空间的存储间，也是自行车的配给系统。它的开发是为了对那些位于普通公共交通枢纽3公里以内的商业区域添加便捷。虽然一开始自动化自行车存储间的目标群体是上班族，但它的简易使用、简单性、适应性和低成本性，使其在多种休闲活动中得到广泛应用。

紧凑而简单

自动化自行车存储间是世界上最为精简的自行车存储系统，其独特的设计使自行车彼此之间的位置仅为16厘米。这使得该系统的成本比其他知名的自动化自行车存储系统低了惊人的70%。

安全性

每个用户的身份会连接到自行车内置的电子识别芯片（RFID）上。这已被证明大量减少了盗窃和人为毁坏方面的问题，与20世纪70年代声名狼藉的阿姆斯特丹的维特自行车方案（又叫"白色自行车方案"）——一个完全匿名的系统——形成鲜明对比。这个成功的方法当前用于OV Fiets（荷兰铁路）和Call-a-Bike（德国联邦铁路）。

环境效益

自动化自行车存储间是利用了自行车作为上班族乘坐交通工具链条中缺失的一环的想法。其核心理念是为那些当前投入使用的不完美的方案创造更好的选择（例如公共交通或汽车）。自动化自行车存储间因使用方便，战胜了消费者不愿使用自行车来替代交通运输的观念。一辆好的自行车的实用性是始终如一的，消除了把折叠式自行车搬到公共交通工具上的现象。自动化自行车存储间占用的空间很小，甚至可以安置于地下。

DESIGNER
David Karásek, Radek Hegmon

DESIGN COMPANY
mmcité

Tyre

轮环自行车停车位

A simple yet attractive shelter with sides inspired by bicycle wheel spokes. Its arched roof made of a trapezoidal steel sheet follows the segment of a giant imaginary wheel circumference and thus perfectly covers parked bicycles.

一个简单但是非常吸引人的遮蔽，侧面的灵感来自于自行车车轮幅度。其拱形防护顶由梯形钢板仿照一个巨大虚拟圆轮的一部分制成，从而完全遮盖住停放的自行车。

PUBLIC RECREATIONAL FACILITY

公共休闲游乐设施

DESIGNER

Jair Straschnow ,Gitte Nygaard

Off-Ground

多功能吊椅

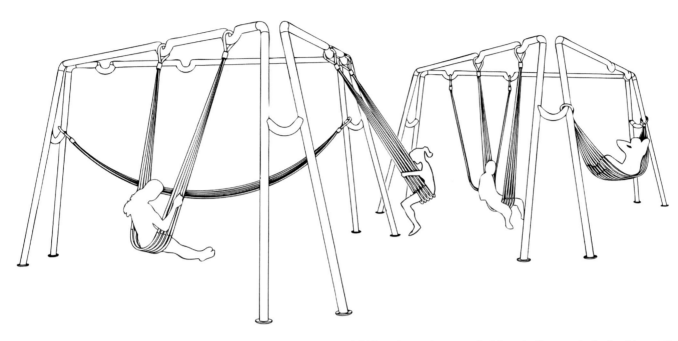

Working assumption #1: Seating facilities in public areas sum up to rigid benches, where comfort is not often a priority (and in most cases the opposite is the case). There are no alternative ways of sitting even in leisure areas and parks, where we'd rather relax and sit more comfortably.

Working assumption #2: Play is free, and is in fact freedom. Play is essential to our well-being. Play is mainly associated with children, playing elements in public space, if provided, are always scaled down to kids' size.

Working assumption #3: Placing things in the public realm is the most democratic incarnation of design: it's free and accessible for all. This is where designers want to operate – this is where real difference can be made.

The designers took the opportunity of a residency period in the Danish Art Workshops during spring 2013, to try and come up with a project that aims to fill what is so obviously missing from our public space.

In short, Off-ground are playful seating elements. Off-ground is about up-scaling playing elements combined with seating alternatives; hanging, floating, swinging, laying - one size fits all. Each seating element can be easily shifted between a low seat, a hammock and a swing, and change according to the user's needs. Sustainability is always integrated in the way of thinking and designing, in this case using rejected fire-hoses as raw material for the seats.

The Off-ground installation shows a different approach to the way public space is perceived and used, basing the design on a true and banal need, but implementing a solution that is functional and playful. Off-ground features this summer in two locations in Copenhagen: at the DAC, Danish Architecture Centre and at Carlsberg City.

Off-ground is the brainchild of award-winning designers Jair Straschnow and Gitte Nygaard.

工作设想 #1：公共区域的休息设施都是些坚硬的长椅，舒适并不是优先考虑的事情（而且在绝大多数情况下，事实往往是舒适度是最被忽略的事情）。甚至在公园与休息区，人们都不知道该坐哪里，而恰恰就是在这时，人们才更想要舒舒服服地坐下来休息一下。

工作设想 #2：游戏是免费的，是一种自由的体现。游戏对我们的幸福有着至关重要的作用。人们总是将游戏与孩子联系在一起，如果在公共空间提供可以游戏的环境，那么通常要将其按比例缩小到适合孩子的大小。

工作设想 #3：在公共领域内放置某些东西：对于所有人来说都是免费且可以使用的。这样的设计是最民主的体现。这就是设计师想要达到的效果——于是就形成了真正意义上的与众不同。

2013 年春天，设计师们在丹麦艺术工作坊居住过一段时间。借此机会，设计师尝试设计一个项目，旨在把公共空间明显缺失的东西填补进去。

简而言之，多功能吊椅创造了一种活泼的休息环境，它将扩大的游戏环境与座椅替代物结合在一起；悬挂、飘浮、摆动、平放——一个尺寸可以满足所有需求。每个设施都可以在低座椅、吊床与秋千之间轻松转换，而且可以根据使用者的要求随意改变。设计师在思考与设计时，可持续性一直贯穿始终。譬如，设计师将废弃的消防水带用作座椅的原材料。

多功能吊椅这一设施显示了公共空间不同于以往的感觉与用途，其设计基于人们真实、平凡的需求，却兼具功能性与趣味性。多功能吊椅成了今年夏天哥本哈根两个地方的特色：丹麦建筑中心与嘉士伯市。见过并使用过多功能吊椅的人的反响进一步地鼓励我们将这一项目带到更多的地方去。

多功能吊椅是获奖设计师 Jair Straschnow 与 Gitte Nygaard 智慧的结晶。

LANDSCAPE ARCHITECT	CLIENT	LOCATION	AREA
LODEWIJK BALJON landscape architects	University of Twente	University of Twente, Enschede, the Netherlands	10,000 m²

Square of Knowledge

知识广场

The Master Plan for the campus of the University of Twente clusters existing and new buildings, rearranges the traffic structure accordingly and creates a new entry. The bases for the new arrangement are the structure of the landscape and the use of the ecological potential of the water management.

The Square of Knowledge is the first step in this development; centred around the square are the buildings for Research & Education. The square is focused on the interaction between students and staff. The space is large and open for events. A tent structure connects the various buildings and gives shelter. The backdrop of trees and evergreen hedges combined with the scattered red polyester benches give ample opportunity for informal meetings in small groups.

In the second phase, the square will be completed, with for instance the extension of the channel. The water is then connected to a deep pond that is part of the cooling system.

屯特大学校园的总体规划中现存建筑与新建筑并存，相应地，该规划对交通路线进行了调整，并开放了一个新入口。

新布局建立在原有景观结构的基础上，并充分利用了具有生态潜能的用水管理方法。

知识广场是本开发项目的第一步；多个研究教育大楼以广场为中心环绕于四周。广场的设计注重学生与员工之间的互动。这里的空间宽敞，适合各种活动。一个帐篷型的结构连接了各个大楼并起到遮阴的作用。红色聚酯长椅点缀在树木和常青树篱组成的绿色背景幕上，为小团体非正式的聚会提供了丰富的空间。

在总体规划的第二阶段，广场会进行一些修整，例如将原有通道进行扩建，从而使广场变得完整。水作为冷却系统的一部分，会被引向一个较深的池塘。

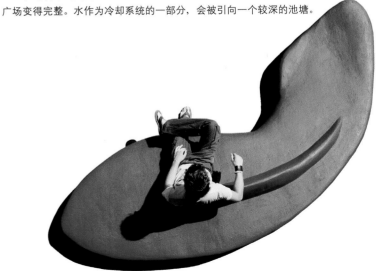

MATERIAL

Gravel, textile, water, custom-made polyester benches, planting (trees, hedges, aquatic plants)

PHOTOGRAPHER

Lodewijk Baljon, Smits Rinsma

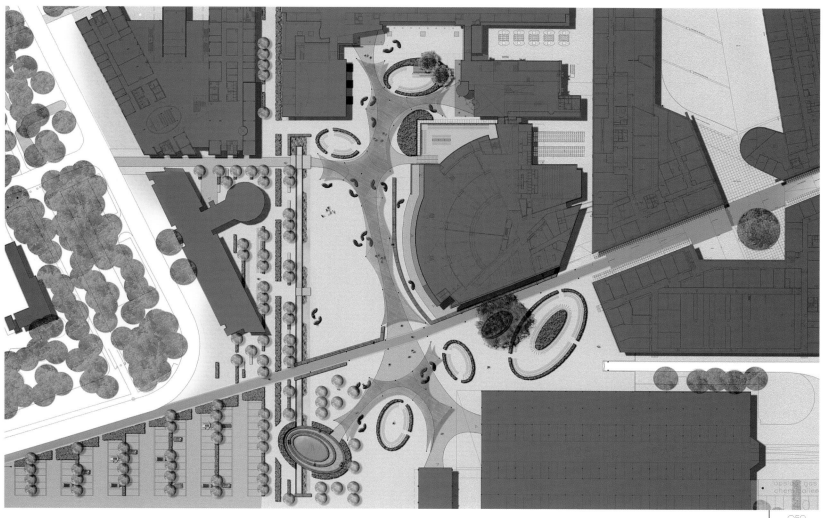

DESIGN COMPANY

Bjarke Ingels Group

ARCHITECT

Bjarke Ingels, Nanna Gyldhom Møller, Mikkel Marcker Stubgaard

COLLABORATION

Topotek1, Superflex, Lemming Eriksson, Help PR & Communication

Superkilen

Superkilen 公园

The conceptual starting point is a division of Superkilen into three zones and colours – green, black and red. The different surfaces and colours are integrated to form new, dynamic surroundings for the everyday objects.

Market/Culture/Sport – the Red Square

In addition to the cultural and sportsfacilities, the Red Square creates the setting for an urban marketplace which attracts visitors every weekend from Copenhagen.

Fitness area, Thai boxing, playground (slide from Chernobyl, Iraqi swings, Indian climbing playground), sound system from Jamaica, a stencil of Salvador Allende, plenty of benches (from Brazil, classic UK cast iron litter bins, Iran and Switzerland), bike stands and a parking area.

Urban Living Room – the Black Square

Mimers Plads is the heart of the Superkilen Masterplan. This is where the locals meet around the Moroccan fountain, the Turkish bench, under the Japanese cherry trees as the extension of the area's patio.

The bike traffic is moved to the east side of the Square by partly solving the problem of height differences towards Midgaardsgade and enable a bike ramp between Hotherplads and the intersecting bike path connection.

Sport/Play – the Green Park

Bauman once said: "Sport is one of the few institutions in society, where people can still agree on the rules." No matter where you're from, what you believe in and which language you speak, you can always play football together. This is why a number of sports facilities are moved to the Green Park, including the existing hockey field with an integrated basketball court as it will create a natural gathering spot for local young people from Mjolnerpark and the adjacent school.

The green park is turning into Mimers Plads on the top of the hill to the south. From the top of the hill you can almost overlook the entire Superkilen.

TEAM Ondrej Tichy, Jonas Lehmann, Rune Hansen, Jan Borgstrøm, Lacin Karaoz, Jonas Barre, Nicklas Antoni Rasch, Gabrielle Nadeau, Jennifer Dahm Petersen, Richard Howis, Fan Zhang, Andreas Castberg, Armen Menendian, Jens Majdal Kaarsholm, Jan Magasanik, Ole Hartmann, Anna Lundquist, Toni Offenberger, Katia Steckemetz, Cristian Bohne, Karoline Liedtke, Jakob Fenger, Rasmus Nielsen, Bjørnstjerne Christiansen

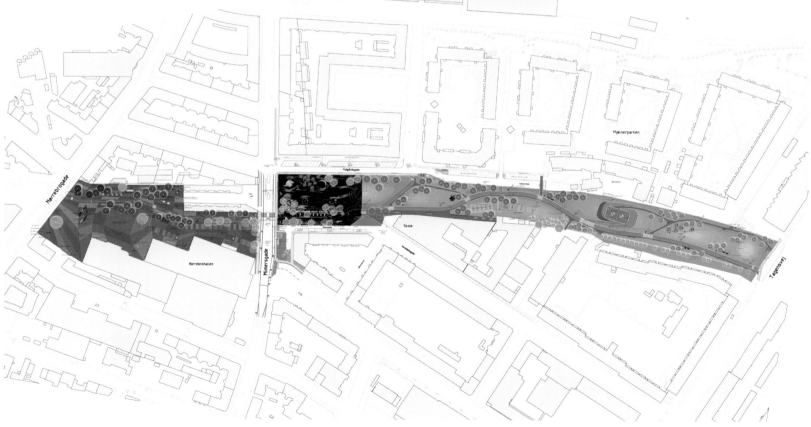

CLIENT

Copenhagen Municipality, Realdania

LOCATION

Nørrebro, Copenhagen

AREA

30.000 m²

构想的起点是将 Superkilen 分成三个区和颜色：绿色、黑色和红色。不同的表面和颜色结合成日常事物的新的、动态的环境。

市场／文化／体育——红色广场

除了文化和健身路径，红色广场创造了城市市场设施，每个周末从哥本哈根和郊区吸引很多游客。

健身区，泰拳区，操场（包括 Chernobyl、伊拉克秋千、印度爬滑梯），来自牙买加的音响系统、萨尔瓦多阿连德模板，许多座椅（来自巴西，经典的英国铸铁垃圾箱，伊朗和瑞士），车站和停车场。

城市客厅——黑色正方形

Mimers Plads 是 Superkilen Masterplan 的心脏地区。在这里，当地居民在摩洛哥喷泉、土耳其台、以日本的樱花树作为扩展的面积的露台等地周围聚会。

自行车交通系统被移到广场的东边，一部分解决了 Midgaardsgade 地区高度不同的问题，也使 hotherplads 和交叉自行车路径有坡道连接其间。

体育／游乐——绿色公园

鲍曼曾经说过："运动场是社会上极少数人仍然相信规则的地方。"不管你来自何方，信仰哪种宗教，说着哪国语言，总能一起来踢一场足球。这就是许多体育设施包括现有的曲棍球场与一个集成的篮球场被转移到绿色公园的原因，它们将为从 Mjolnerpark 地区和附近学校前来的当地青年人创造一个天然的聚集地。

DESIGNER
Marcus Abrahamsson

DESIGN COMPANY
Studio MA.A&D

Bikers Rest

骑车人休息架

The Designer and Architect Marcus Abrahamsson has for Swedish Nola designed a new kind of bollard for the rapidly growing number of bicycle commuters all around the world. He saw other bicyclists constantly trying to time the green light, slowing down towards the stop, turning their wheels to keep balance and eventually giving up and in defeat getting down of the saddle. This bollard gives cyclists a footrest to lean against and a handle to hold on to while waiting for the light to change.

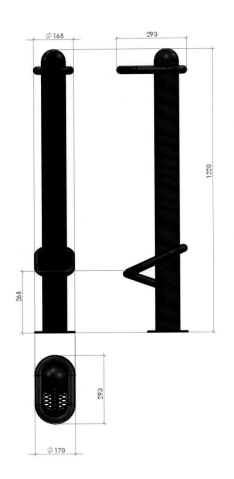

　　全世界每天骑自行车上下班的队伍在不断壮大，针对这一情况，身为设计师与建筑师的马库斯·亚伯拉罕松为瑞典诺拉户外家具公司设计了一种新型短柱。他看到那些骑自行车的人总是在不断地尝试根据绿灯设定车速，减速骑向规定的停止指示线处，还要转动车轮保持平衡，但最终不得不放弃，垂头丧气地从车座上下来，因此产生了灵感。这种短柱上有一个脚踏板和一个把手，这样一来，骑自行车的人们就可以踏着脚踏板、抓着把手来等候信号灯改变颜色。

DESIGNER	DESIGN COMPANY	MATERIAL
Mouna Andraos, Melissa Mongiat	Daily tous les jours	Steel, rubber, sound system, motion capture system, LED lights, computer, soundtrack

21 Balancoires

音乐秋千

21 Balançoires is a giant collective instrument, a game where together we achieve better things than separately.

When in motion, each of the 21 swings in the series triggers different notes and, when used all together, the swings compose a musical piece in which certain melodies emerge only through cooperation.

Together with Luc-Alain Giraldeau, an animal behaviour professor from the Universite du Quebec, a Montreal's Science Faculty, DTLJ explored the concept of cooperation:

Cooperation emerges when the behaviour of each individual depends on the decisions of the rest of the group: it's a game where, from the start, you need to adjust to the actions of others.

The result is that stimulates ownership of the space, bringing together people of all ages and backgrounds, and creating a place for playing and hanging out in the middle of the city centre.

DIMENSIONS	LOCATION	PHOTOGRAPHER
H 3.3 m x L 123 m x W 5.5 m	Montreal, Canada	Olivier Blouin

音乐秋千是一个大型的合成器材，比起单独游戏，它更是一个我们大家一起去做的游戏。

有秋千运动时，该系列中的每个秋千会触发不同的音符，当所有的秋千一起运动时，就会谱成一首音乐作品，只有通过合作才能奏出特定的旋律。

Luc-Alain Giraldeau 是魁北克蒙特利尔综合理工大学研究动物行为的教授，与他一起探讨合作的理念：当个体的行动由团体其余人去决定时，才会出现合作——这是一个游戏，从一开始，你就需要去配合别人的动作。

该装置刺激了当地人的归属感，使各个年龄层和不同背景的人聚集在一起，在市中心创造了一个玩耍和闲逛的天地。

Swing

公共秋千

Swing is an interactive and playful installation conceived by Moradavaga for the Pop Up Culture programme promoted by Guimarães 2012 – European Capital of Culture. Based on the principle of swinging to produce electricity, Swing is also an ode to the rich industrial heritage of Guimarães, reflected in its mechanical devices and sounds evocative of the ones once produced in the factories of the city. The base structure, made of re-usable wooden pallets, serves as a podium for the swings at the same time that it contains the hidden electrical system, making it a surprising experience for the users when they perceive the lights turning on beneath their feet as they start swinging. Traditional hemp rope, wooden beams, bicycle chains, wheels, dynamos and lights complete the material palette used in the installation giving it an "old-style" look and a low-tech kind of feel.

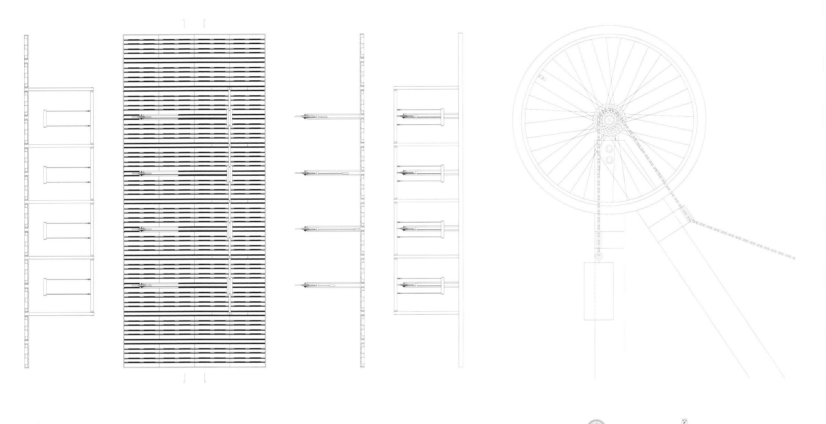

DIMENSION

1165 cm long x 485 cm wide x 235 cm high

PHOTOGRAPHER

Manfred Eccli & Pedro Cavaco Leitão

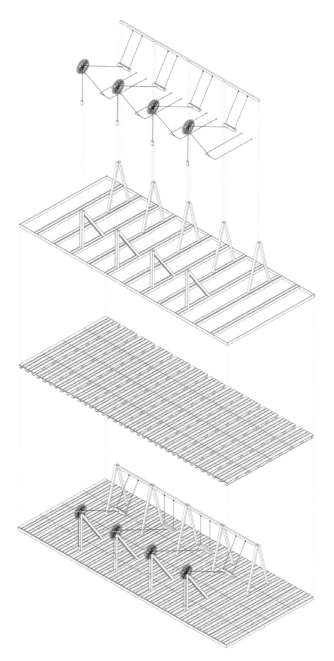

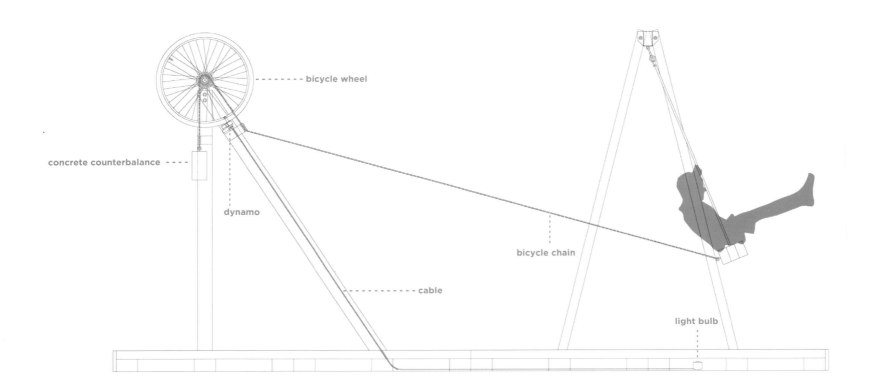

bicycle wheel

concrete counterbalance

dynamo

cable

bicycle chain

light bulb

公共秋千是一项互动性的游乐设施，由设计公司 Moradavaga 构思，为 Guimarães 2012——欧洲文化首都的文化节目所设计。在摆动的原则基础上来产生电力，公共秋千是丰富的工业遗迹的颂歌，反映了它的机械设备和机器一度回荡在城市的声响。用循环使用的木板做成的底座为秋千提供了一个平台，同时它也包含着隐藏的电力系统，当使用者开始摇摆，他们会发现在地台下的灯具打开了，这给他们一种奇妙的体验。传统的麻绳、木梁、自行车链、车轮、发电机和灯，完成了设施的材料组合，给秋千一个"老式"风格的外观和低技术含量的感觉。

DESIGN COMPANY	STRUCTURAL DESIGN	PRODUCTION	LOCATION	PHOTOGRAPHER
TORAFU ARCHITECTS	Ohno Japan	Ishimaru	Tokyo, Japan	Fuminari Yoshitsugu / TORAFU ARCHITECTS

Gulliver Table

巨长桌

The Midtown DESIGN TOUCH Park kicked-off in a wide open grass field in Tokyo Midtown where design can be enjoyed with the five senses. Here, TORAFU ARCHITECTS proposed an open stage on which children and adults alike take part in workshops while having fun.

A very long table has suddenly appeared on the lawn. They installed a tabletop running 50 metres in a straight line, as if drawing a horizontal line on the gently sloping grassy field.

The table naturally grows taller as we move along the inclined plane. The structure serves as a symbol for the children's workshop – it is a floor, a table, a bench, a playground or even a shelter.

Various persons come all together around a single 50-metre-long table, discovering various places along the way; places to sit down, to stand up, to lie down or pass under.

Like a big family gathered around a picnic table, this is a stage full of smiles where all citizens, young and old, converge to create something new.

这张大桌子由 Torafu 建筑公司所设计，放置在东京中心城区公园的一块开阔的草坪上，让公众从感官上切实地感受到它。该公司设计的巨长桌让孩子和大人可以在这里享用桌子的同时，还获得了乐趣。

如此长的桌子突然出现，为草坪增添了一份乐趣。巨长桌的桌面长 50 米，直线形斜放着，仿佛是在平缓的草坪上画出的一条地平线。

当我们沿着桌边行走时，会发现它在逐渐地"长"高。对孩子来说，这样的建筑可以是一层木板、一张桌子、一张长椅、一个游乐场，甚至是一个堡垒。老人、年轻人、儿童都能在这张长 50 米桌子的不同位置找到自己喜欢的享受方式，或坐下、站着、躺下，抑或行走于其下。

犹如一家人在一起聚餐玩耍一般，这是一个让市民们，不论老少，充满欢乐的地方，提高了该区域的温馨感。

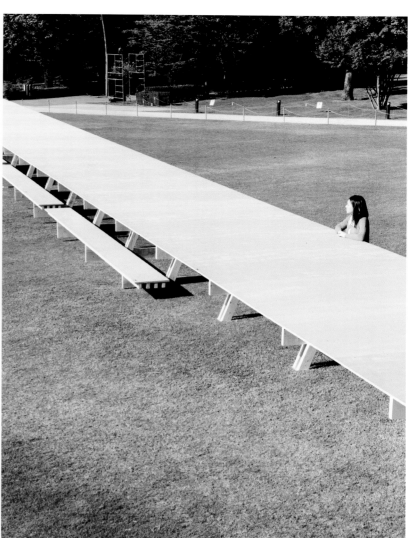

Joao Meireles Park Avenue

若昂・梅勒雷斯公园大道

The project is part of a wide intervention that spreads from this central axis to the rest of the city. The main goal was to transform an uncharacteristic space in a qualified and humanised axis. The strategy has implications on wide urban flows, creating roots to positive urban growth and regeneration. The aim is to boost circulation and use in a multi-programmatic narrative, where green is a constant, focusing on the well-being and full access to useful services, sports facilities or leisure equipment. The programme suits various needs, result of a part buoyant, part resident, population. The concept is based on the theme of a continuous path that crosses a structural green axis, where elements appear as programmatic and iconographic moments, inform of platforms, characterising a positive city personality. The work reached to all possible variables of intervention, from landscaping to public lighting or the design of the urban equipment, integrated in the platforms and distributed in the circuit.

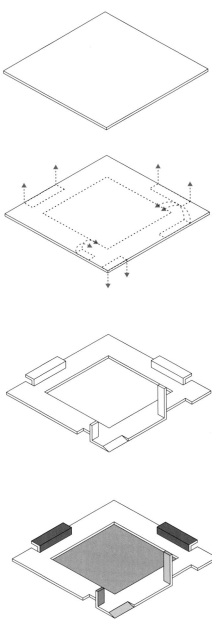

DIMENSION
Park length: 1,400 m

LOCATION
Vilamoura, Portugal

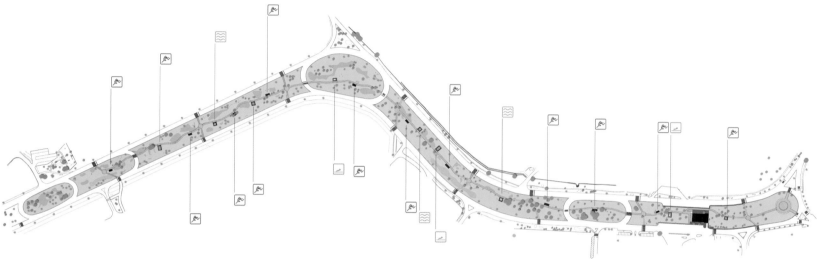

GLOBAL PROGRAM:

// 1400 meters of intervention
// 8 transversal connections (3 bridges)
// 16 leisure/sports platforms

11x   sport platform
03x   lounge platform
02x   water platform

// designed urban furniture equipment:
trash cans / information plates / drinking fountains / benches / global lighting

　　该项目是大型工程的一部分，从中轴线一直延伸到该市的其他地方。项目主要目标是要将一个没有典型特征的空间改造成合格的、人性化的中心枢纽。该策略对城市大量人口流动产生影响，为城市的积极发展与新生创造了条件。开发中轴线目的是让人们更多地来到这里，享受多个开发项目的福利，这里有绵延的绿化带，让市民更健康，并充分利用服务体系、体育设施，或者休闲设施来满足需求。该项目适应了不同人群的需求：既满足了部分流动人群，又服务了附近居民。建设的理念是要打造一条绵延的绿色主干道，两旁分列各种陈设和设施、平台等，塑造出这个城市的积极风貌。项目工程的开展会涉及方方面面，从美化景观到公共照明，或者城市设施的设计、集成平台和电路分布等。

PAVEMENT
地面铺设

The Pool | Place Emilie-Gamelin
泳池｜埃米丽－干莫林广场

Schoolyards Various Localities, Quebec, Canada
加拿大魁北克各地校园运动场

Levinson Plaza, Mission Park
莱文森广场，米森公园

Arena Boulevard & Amsterdamse Poort
竞技场大道与阿姆斯特丹大门

Kømagergade
Kømagergade 购物街

Damsterdiep
Damsterdiep 路面

DESIGN COMPANY

NIPpaysage Landscape Architects

TEAM Michel Langevin, Mélanie Mignault, Josée Labelle, Mathieu Casavant, Sylvain Lenoir, Emilie Bertrand-Villemure, Claude Cournoyer, Georges-Étienne Parent.

The Pool | Place Emilie-Gamelin

泳池｜埃米丽－干莫林广场

Public swimming pools in Montreal are little islands of life, movement and refreshment. They are the theatre of summer, a place for relaxation and play in contrast to Montreal's hot and humid urban reality. In summer 2011, a virtual swimming pool was set up at Place Emilie-Gamelin, a large public plaza in the centre of the city. Equipped with playful furniture reminiscent of water fun, the swimming pool hides in its pixels an enormous hopscotch court, numerous chess boards and a giant tangram puzzle. The installation is designed to accommodate large crowds during festivals and special events including the park summer programme organised by Le partenariat du Quartier des Spectacles.

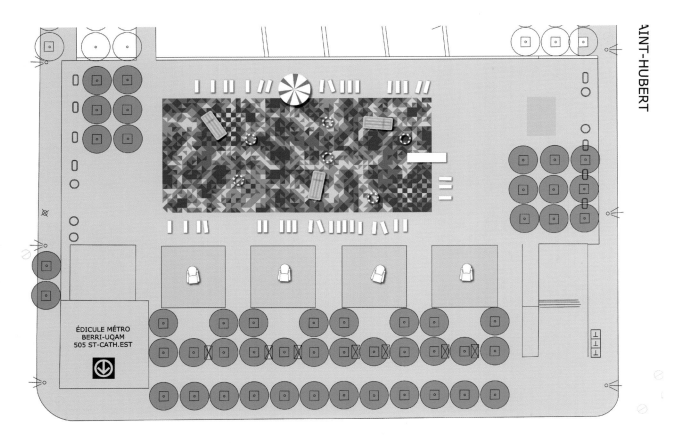

CLIENT	AREA	LOCATION	PHOTOGRAPHER
Partenariat du Quartier des Spectacles	2,470 m²	Montreal(Quebec), Canada	Frédérique Ménard-Aubin, NIPpaysage, Vincent Lemay

　　蒙特利尔的公共泳池就如同人们生活、运动和休闲的小岛。在炎热潮湿的夏日里，公共泳池就是娱乐中心，人们可以在那里放松和娱乐。在 2011 年的夏天，埃米丽莫甘林广场上建起了一个虚拟游泳池，在城市中心形成了一个大型公共广场。广场配备了有趣的设施，使人联想起戏水的欢乐。游泳池隐藏在地面铺设的像素图案中，其中包括一个巨大的跳房子游戏场地，多个国际象棋棋盘和一个大型的七巧板图案。这样的设计可以在节日和特别活动时，包括 Le partenariat du Quartier des Spectacles 举办的公园暑期项目时，容纳大量人群。

LANDSCAPE ARCHITECT	CLIENT	LOCATION
NIPpaysage Landscape Architects	Various schools and school boards	Montreal (Quebec), Canada

Schoolyards Various Localities,Quebec,Canada

加拿大魁北克各地校园运动场

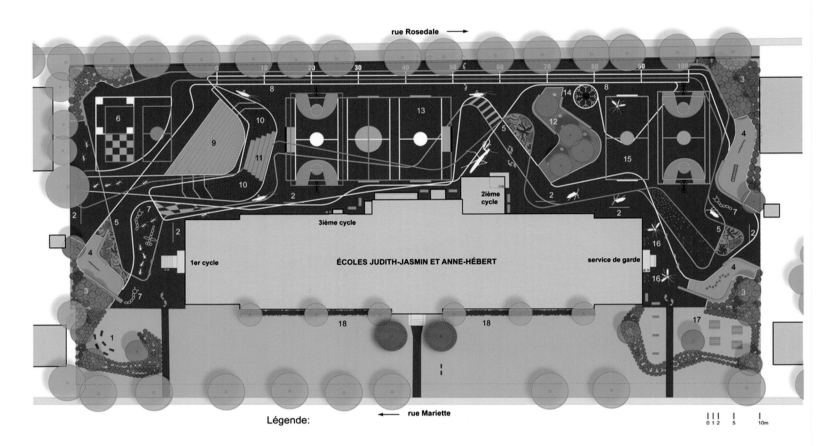

1. reading area under the tree
2. punching bag structur
3. green island
4. thematic island with equipment
5. stage-medidation area
6. 1st cycle
7. hopscotch game
8. 100m path-
9. synthetic grass slope
10. asphalt berm
11. wood amphitheatre

Case 1 – Paul-Bruchési Schoolyard

Originally entitled "Blue Yard", the project was built with a very limited budget. For economic reasons, the school decided to keep the existing asphalt paving and build a new project on top. NIPpaysage decided to work within this requirement by exploring large patches of painted colours and softer surfaces to give the site a well-deserved landscape makeover. The result was achieved by implementing a strong coloured scheme to organise and structure different uses and programmed activities. Using circulation patterns and building volumes, a grid of white lines was established, defining and organising play and other leisure spaces.

Case 2 – St-Enfant-Jésus Schoolyard

The school offers specialised services to a clientele of hearing and/or visually impaired students. To address specific needs, NIPpaysage proposed a rationalisation of the larger paved space by combining play, imagination and security objectives into an overall design scheme.

Case 3 – St-Paul-de-la-Croix Schoolyard

By using the existing topography as a dynamic activity anchor, destination and viewing point, NIPpaysage created diverse opportunities for school activities, including relaxation, informal and structured play. To emphasise the importance of sports and student movement through the site, the schoolyard landscape is circumscribed by a zigzagging contemporary running track that follows and unifies the two main levels of the schoolyard space.

TEAM   Mathieu Casavant, Josée Labelle, Michel Langevin, Mélanie Mignault, Emilie Bertrand-Villemure, France Cormier, Claude Cournoyer, Suzanne Ernst, Georges-Étienne Parent, Céline Paradis, Gabriel Mauchamp, Mélissa Marcotte, Yoanne Source, Sylvane Rava, Baptiste Piednoël, Delphine Dalençon, Patrick Morand, David Lazcano

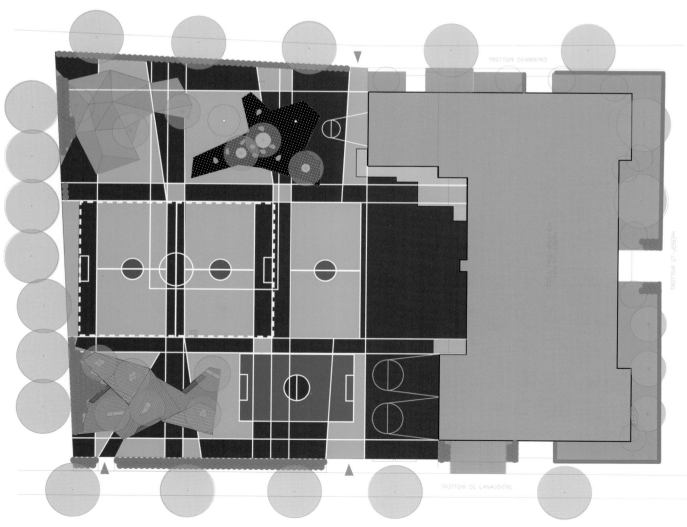

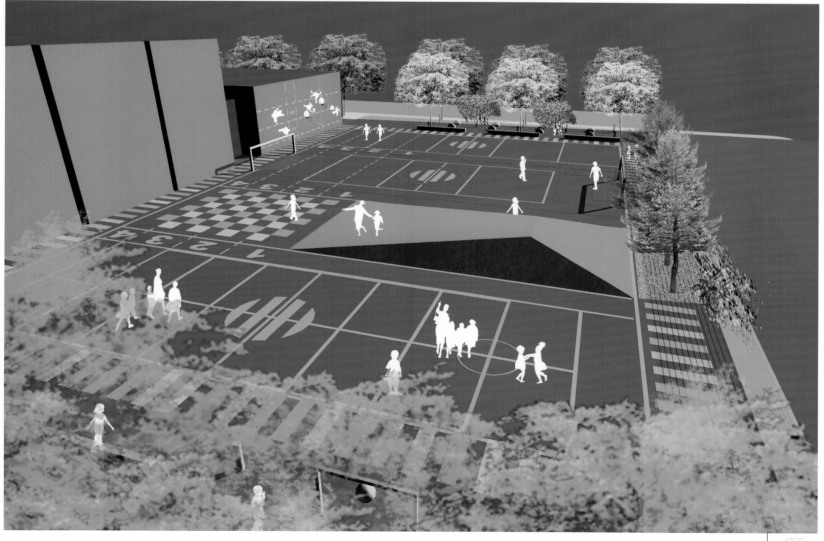

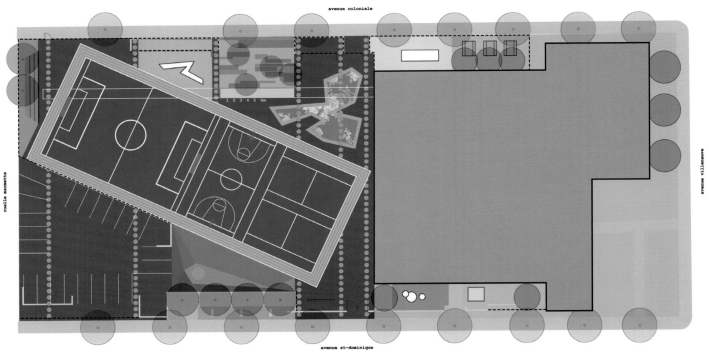

案例一：PAUL-BRUCHÉSI 校园运动场

该项目起初被称为"蓝色之地"，建筑预算十分有限。由于经济原因，学校要求保留原有的沥青路面，在其基础上建设新项目。建筑师 NIPpaysage 根据这一要求展开工作，通过采用大量手绘彩色补块和柔软表面对该地区的景色进行彻底的改造。他运用高饱和色彩组织方案，建造不同用途和规划的活动区，使该建筑效果得以实现。

运用循环图案在建筑之间画出一道道纵横的白线，以此规划出游戏区和其他休闲区域。

案例二：ST-ENFANT-JÉSUS 校园运动场

该学校为听力或／和视觉有障碍的学生提供了专业化的服务。为了满足特殊的需求，NIPpaysage 提出了一个设计方案，将娱乐、想象与安全目标融入整体设计方案中，使大片铺设空间变得合理化。

案例三：ST-PAUL-DE-LA-CROIX 校园运动场

NIPpaysage 利用现有地势作为动态活动的支撑点、目的地和观测点，为学校活动创造出各种各样的机会，包括休闲活动、非正式而有组织的活动。为了强调运动的重要性以及方便学生在这里活动，操场景观被蜿蜒的跑道围了起来，跑道将校园运动场不同的两个平面连接起来。

DESIGN COMPANY	LOCATION	AREA	PHOTOGRAPHER
Mikyoung Kim Design	Boston, Massachusetts	2,787 m²	Lisa Garrity, Charles Mayer

Levinson Plaza, Mission Park

莱文森广场，米森公园

The plaza landscape design focuses on providing Mission Park with a landscape which draws its spirit from the regional gardens of New England. The plaza uses pavement materials which will endure the long, challenging winters, while the patterning itself is designed based on the herringbone patterns of residential landscapes. The plaza pattern stitches the areas of gathering and passage together while bringing a human scale to this large plaza space. The paved areas within the garden are carved out of the grove to allow for direct access to major entry points and public transportation.

　　广场景观设计的重点是给米森公园提供景观，灵感来自新英格兰区的公园。该广场采用铺面材料，它能忍耐漫长并具挑战的冬天，而这个图案本身是基于住宅景观的人字形图案的。广场的图案把聚集地和通道的区域拼接在一起，使这个大型的广场空间具备了个人的范围空间。公园内铺好的区域分割出了一片片小树林，可以从主入口通向公共交叉点。

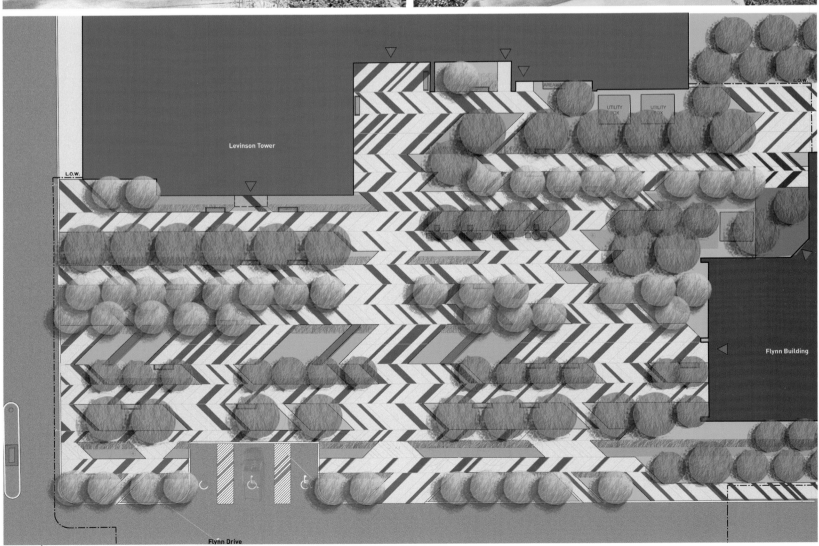

Arenaboulevard & Amsterdamse Poort

竞技场大道与阿姆斯特丹大门

The Arena Boulevard and the Amsterdamse Poort will together be developed into Amsterdam's second nightlife district. The current central area has two different faces: the busy, small-scale Amsterdamse Poort shopping centre and the spacious but often empty Arena Boulevard. The buildings, in which major functions are accommodated behind a small number of entrances, accentuate the impersonal character of the area. The Aerena Boulevard is everything that historic city centres are not. The emphasis in the design for the Arena Boulevard is on breaking up its lianre character, and creating a space that is pleasant for a group of ten people, but also for a crowd of fifty thousand.

Transverse connections between the north and south side are enhanced by creating space between the buildings on the boulevardn. The design of the public space organises itself in a natural way. Long benches of natural stone and wood mark the transition between places for movement and places to pause. Alongside these benches the ground level rises up or falls away, giving a slightly curved paved surface. Some points are laid out for sports use, while others are raised up, with trees and grass creating places to relax. A public space is created that invites a longer stay and offers opportunities for various kinds of use.

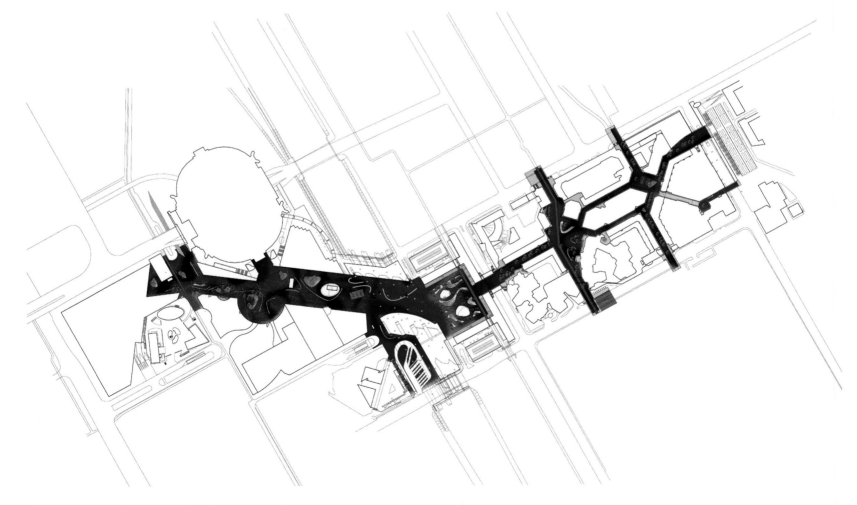

ASSOCIATE	CLIENT	LOCATION	AREA
ARUP Lighting	Municipality of Amsterdam	Amsterdam, The Netherlands	42,000 m²

竞技场大道和阿姆斯特丹大门将一起成为阿姆斯特丹的第二个夜生活区。目前中心区呈现两种不同的面貌：忙碌但规模小的阿姆斯特丹大门购物中心和广阔却经常空无一人的空竞技场大道。少数的入口后面的建筑，提供了主要功能，强调了可观区域的特征。竞技场大道是一切历史悠久性城市的中心。竞技场大道设计的重点是打破了它的线性特征，并创造一个场所，十人一组是惬意的，但同时也能挤满五万大众。

北区和南区之间的贯轴连接，通过创建林荫大道上的建筑物之间的空间得到了增强。公共场所的设计以一种自然的方式组织起来。用天然石材和木材制作的长凳，标记着运动场所和休息区域之间的过渡。沿着这些长椅，地面高度上升或下降，产生一种略微弯曲的铺面。一些地点铺设为体育用地，其他地方有所提高，树木和小草营造了放松的地方。公共场所被建造来用于邀请人们更长时间地停留和提供使用各种公共空间的机会。

DESIGNER
KBP.EU, a joint venture of Karres en Brands and Polyform

DESIGN COMPANY
Karres en Brands landscape architecture + urban planning

Kømagergade

Kømagergade 购物街

The curved course of the Kømagergade shopping street is typical of the city centre of Copenhagen. The redesign of the public space is called Købmagergade, which consists of the main shopping street (Købmagergade) and three squares (Hauser Plads, Kultorvet and Trinitatis Kirkeplads). The Købmagergade shopping street embodies the characteristic image of the labyrinthine medieval city centre.

The district has its own daily and weekly rhythms: people cycle, walk, shop, play and go out in the evenings, but traffic for deliveries, refuse collection and maintenance also joins in this rhythm. Their first step was to clean and empty the area, so that the flow of people can easily find its way. They also selected strong materials such as natural stone: a durable material with a strong and harmonious appearance. The design proposal encourages the development of intensive city life on the one hand, and on the other it is linked with the rich history of Copenhagen.

The layout of the three squares is varied, just as their historical situation and their location in the city are varied. On the Kultorvet the dark – almost black – paving pattern of the stone is inspired by the 18th-century coal trade. On the rather more peaceful Hauser Plads square, the exciting grass play mounds form a green oasis in the urban fabric. At night, the Trinitatis Church square with its famous observatory Rundetårn is transformed by artificial lighting into an enormous starry sky. The three squares are diverse in colour, from dark coal to bright stars: "From Kultorvet to the Milky Way".

In the evening and at night the medieval city has its own melancholy and mysterious atmosphere, especially in winter. This unique ambiance is emphasised by the use of warm indirect lighting, with a few extra accents on the squares. This means that it is still possible to see the stars, just as Christian the Fourth did from the observatory in Rundetårn in the 17th century.

ASSOCIATE
Oluf Jørgensen Engineering and Ulrike Brandi Licht

CLIENT
Municipality of Copenhagen

LOCATION
Copenhagen, Denmark

AREA
22,000 m²

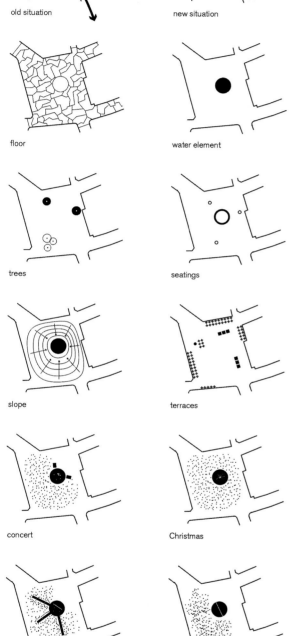

old situation → new situation

floor

water element

trees

seatings

slope

terraces

concert

Christmas

Copenhagen fashion week

open air film

jumble sale

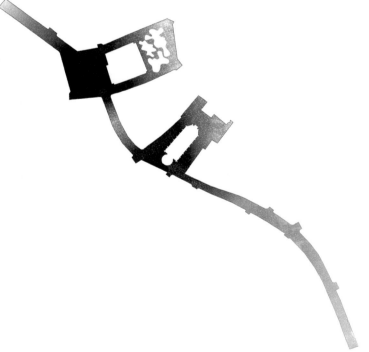

这条弧形的 Købmagergade 购物街是哥本哈根市中心的象征。该公共区域的改造被命名为 Købmagergade，其中包括一条主要的购物街（Købmagergade）和三个广场（Hauser Plads, Kultorvet and Trinitatis Kirkeplads），The Købmagergade 购物街再现了中世纪城市中心的迷宫式特点。

这个区域有其每日和每周的节奏：人们在其中骑自行车、散步、购物、游玩，并在傍晚时外出。物流交通、垃圾收集和日常维护也加入这个节奏之中。他们的第一步是清洁并排空整个区域，使流动的人群可以很容易地找到它的路线。他们还选择了结实的材料，如天然石材，一种坚固而外观协调的耐用的材料。该设计方案一方面鼓励集中的城市生活的发展，另一方面也与哥本哈根丰富的历史相联系。

三座广场的布局各不相同，就像它们的历史地位和在城市中的地位各不相同一样。

在 Kultorvet 广场上，暗色——多为黑色——的石质铺地图案是受到了 18 世纪煤炭贸易的启发。在更为平静的 Hauser Plads 广场上，令人情绪高涨的草地游乐小丘形成了城市建筑中的一块绿洲。在夜间，Trinitatis 教堂广场和著名的 Rundetårn 望台通过人工照明转化成了一片巨大而繁星满天的夜空。这三座广场在颜色上各不相同，从暗沉的煤炭到闪亮的星星，即"从 Kultorvet 到银河"。

从傍晚到夜间，这座中世纪的古老城市有着独特的忧郁而神秘的氛围，在冬季尤为突出。这种独特的氛围通过在广场中使用暖色调间接照明和少量的强调性装饰而得到了进一步加强。这意味着我们仍然可以看到星星，就像克里斯蒂安四世在 17 世纪时在 Rundetårn 望台中所做的那样。

Damsterdiep

Damsterdiep 路面

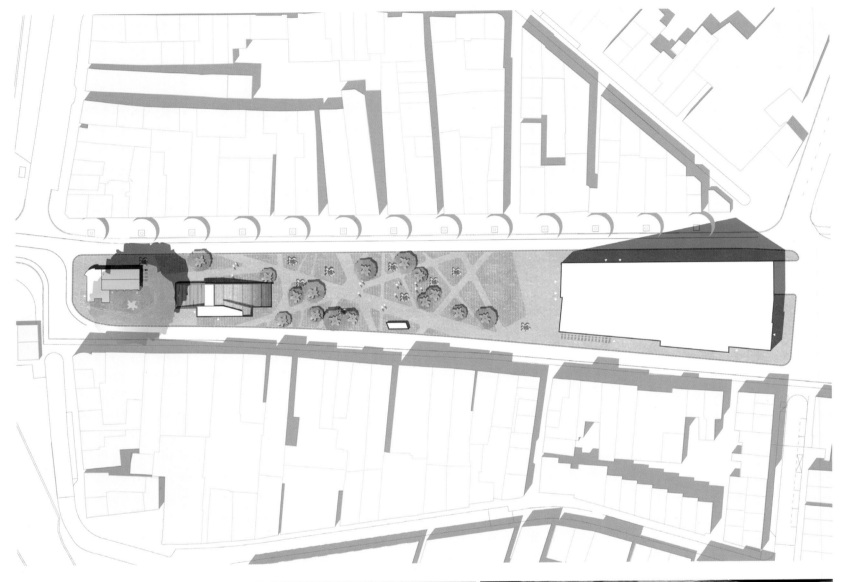

ASSIGNMENT	CLIENT	LOCATION	AREA
public space design	Municipality of Groningen	Groningen, The Netherlands	11,000 m²

The Damsterdiep lies in the inner city of Groningen, on the old course of the canal between Groningen and Delfzijl. The coming of an office building and underground parking garage for 550 cars makes a new approach for this central location necessary.

The well-worn walking routes from entrances, stairways of buildings and surrounding lanes are delineated by matt paths in the polished paving. The pattern of these walking lines breaks the scale and height of the space. A field of dynamic LED-lighting is integrated in the square's paving. In the evenings, moving graphic patterns light up in the ground, referencing the water from the underlying Damsterdiep.

Damsterdiep 路面位于格罗宁根城市内，在格罗宁根和代尔夫宰尔之间的运河故道上。新建的写字楼和 550 个车位的地下停车场为这一中央位置的必要性进行了新的探讨。

入口、楼梯的建筑和周边车道的陈旧的人行路线，被磨砂表面道路在抛光路面划出痕迹。人行道上的图案打断了空间的规模和高度。成片的动态 LED 灯集成在广场的铺路上。到了傍晚，移动的图案在地上亮起来，在 Damsterdiep 路面的下方引用水流。

SEAT
座椅

DESIGNER
Studio Weave

PROJECT INITIATORS
Jane Wood and Sophie Murray

STRUCTURAL ENGINEERS
Adams Kara Taylor

PROJECT MANAGERS AND QUANTITY SURVEYORS
Jackson Coles

MAIN CONTRACTORS
Millimetre Ltd

CLIENT
Arun District Council

LOCATION
Littlehampton, West Sussex

The Longest Bench

最长座椅

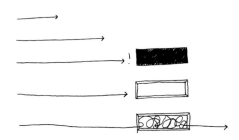

The Longest Bench in Britain was opened to the public in Littlehampton, West Sussex on the 30th July 2010.

The bench seats over 300 people along Littlehampton's promenade, overlooking the town's award-winning Blue Flag beach. Designed by Studio Weave, the structure sinuously travels along the promenade, meandering around lampposts, bending behind bins, and ducking down into the ground to allow access between the beach and the Green. Like a seaside boardwalk the Longest Bench rests gently on its habitat and adapts to its surroundings while like a charm bracelet it connects and defines the promenade as a whole, underlining it as a collection of special places that can be added to throughout its lifetime.

The bench is made from thousands of tropical hardwood slats engraved with messages from its supporters. The timber is 100% reclaimed from sources including old seaside groynes and rescued from landfill. The beautiful variety of reclaimed timbers are interspersed with splashes of bright colour wherever the bench wiggles, bends or dips.

Accompanying the long bench are two bronze-finished steel monocoque loops that connect the promenade with the green behind it. As the bench arrives inside the twisting loops it goes a little bit haywire, bouncing of the walls and ceiling creating seats and openings. The loop contains the haywire stretch of bench and frames the views each way.

Notes:

* The completed phase one structure measures 324m (following the curves of the structure) and seats over 300 adults.

* The full length target would reach 621m (following the curves of the structure), seating over 800 adults.

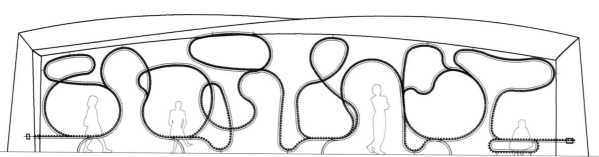

Elevation_of_Shelter

0.5 1 2 5m

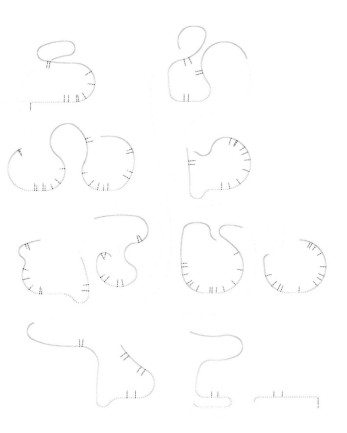

Exploded_Elevation_of_Squiggle

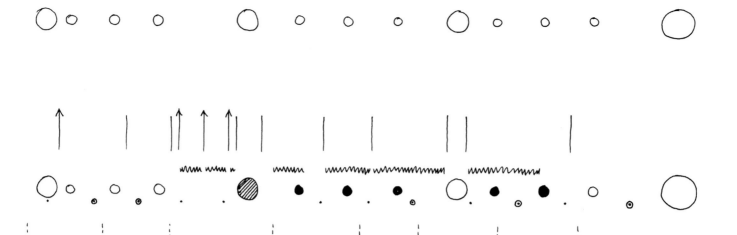

west style   east style

LongestBench_Bead_Diagram

最长座椅设施于 2010 年 7 月 30 号在西萨塞克斯郡的利特尔汉普顿对公众开放。

可提供给超过 300 个人坐的座椅沿着利特尔汉普顿散步走廊，远眺城镇颇受赞誉的蓝旗海滩。座椅由设计师 Studio Weave 设计，它的结构沿着走廊蔓延前行，在路灯旁的铁箱中迂回弯曲，深入地下来确保海滩和绿化带之间通道的连贯。

类似海边木板路的造型最长座椅轻轻地搭建融合于周围的环境，就像一个富有魔力的手镯从整体上连接并定义了长廊，强调了它作为一个很多特殊地点的集合体，可以不断增加新的地点。

座椅用成千上万的热带硬木长木条制成，在木条上面刻有它支持者的名字。木材选用 100% 的再生能源，包括旧的海边防洪堤和垃圾掩埋场的二手建材。板凳扭动、弯曲或下降时反射出的鲜艳颜色上缀着美丽的各种再生木料。

伴随着长椅是两个青铜制成的钢铁回路，连接了长廊和它背后的绿野。当座椅延展至缠绕的环形，它就会变得有些缭乱，在墙壁和天花板上的剧烈起伏塑造了椅座和通道。弧形沿着座椅蜿蜒而出，可以欣赏道路两旁的风景。

注意：

完成第一阶段的结构措施 324 米（以下结构的曲线），可容纳超过 300 人。

全长度的目标将达到 621 米（以下结构的曲线），容纳超过 800 名成年人。

DESIGNER
Dejan Mitov, Jelena Čobanović, Krsto Radovanović and Maja Nogavica

DESIGN COMPANY
ModelArt Studio

Bench 1000cm

十米长椅

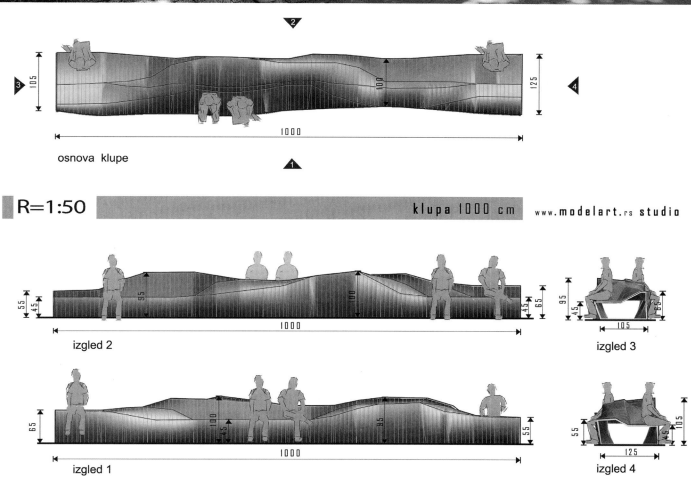

osnova klupe

R=1:50   klupa 1000 cm   www.**modelart**.rs **studio**

izgled 2

izgled 3

izgled 1

izgled 4

"Bench 1000cm" is the new project that has been created by "ModelArt Studio" from Serbia in collaboration with Faculty of Technical Sciences in Novi Sad and the Municipality of Dimitrovgrad. The project was created as a part of research of the influence of architectural installations such as microurban intervention in public space and its revival through the creation of benchmark pointas places of socialisation. It is an experimental spatial installations / street furniture, primarily intended for open public spaces as a utilitarian object, but also it is specific spatial seal that visually dominates the area and makes specail atmopshere for socialising.

The form object is achieved by parametric computer modelling, and then translated into physical form. The structure was made of wooden slats, with alternating toggled full and empty places which were combined with internal lighting that gives an interesting light effects. The structure is 10m long, but it gives possibility of reducing or extension depends on the place where it stands.

"十米长椅"是新的项目，它是由来自塞尔维亚的"ModelArt Arhitekti"与诺维萨德和季米特洛夫市当局技术科学院合作创建的。该项目是作为建筑设施影响研究的一部分创建的，比如在公共场所的微城市干预以及通过创造长椅景观增进社交的场合的使用。这是一个实验性的立体设施／街道景观，主要作为一个实用的设施置于公共开放场所，但同时，它也是一个特定的立体标志，它成为了这个区域的视觉亮点，为人们的社交营造了特有的气氛。

这个形状的长椅用参数化计算机建模做成，然后才转换为物质的外观。这个构造是由木条搭建而成的，它们凹凸有致地交替拴牢，长椅内部结合了内部照明，展示了一个有趣的灯光效果。这张长椅有 10 米长，但它提供的凹陷或突起是由它所置的场地决定的。

DESIGNER	DESIGN COMPANY	MATERIAL
Margus Triibmann	Keha3	recycled plastic

Klimp

面团椅子

Klimp (dumpling) is a funny, convenient, flexible piece of furniture carrying a modern way of thinking and created for public places which at the same time meets all conditions set for a city space whether following the climate, vandalism or comfort of use.

In one of its main positions, Klimp resembles a traditional park bench or a sofa at home. Rotating the bench creates new and unexpected forms comfortable for sitting, leaning or just relaxing.

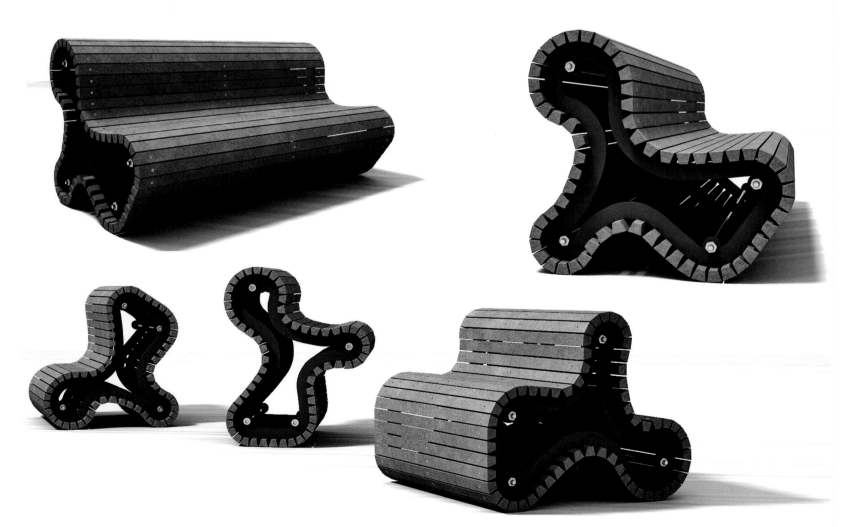

Klimp（面团椅子）是一个有趣、方便、灵活的公共设施，设计师用现代的思维方式为公共场所而创作。不管天气、人为破坏或者使用舒适度如何，它都能以各个方向放置于城市空间中。

把面团椅子的某个椅面朝上，就类似于公园里传统的长椅，或者家中的沙发。翻动这个面团椅子，就能得出新的意想不到的座位造型，提供了舒适的坐席、依靠，或者只是放松一下。

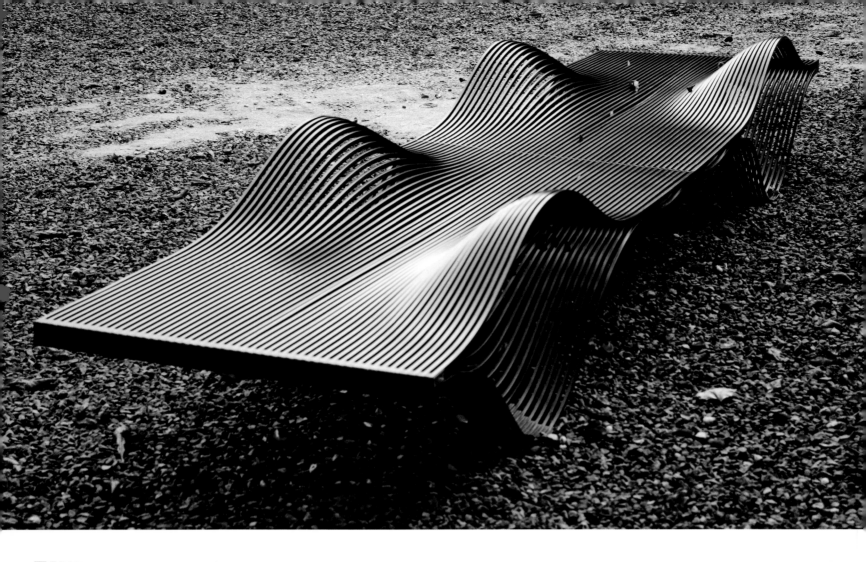

DESIGNER	MANUFACTURER	MATERIALS	DIMENSIONS	TOTAL WEIGHT
Alexandre Moronnoz	TF création	steel 8mm, primary zinc epoxy powder coating	400cm × 78.5cm × 54cm	450kg

Muscle Bench

动感公共座椅

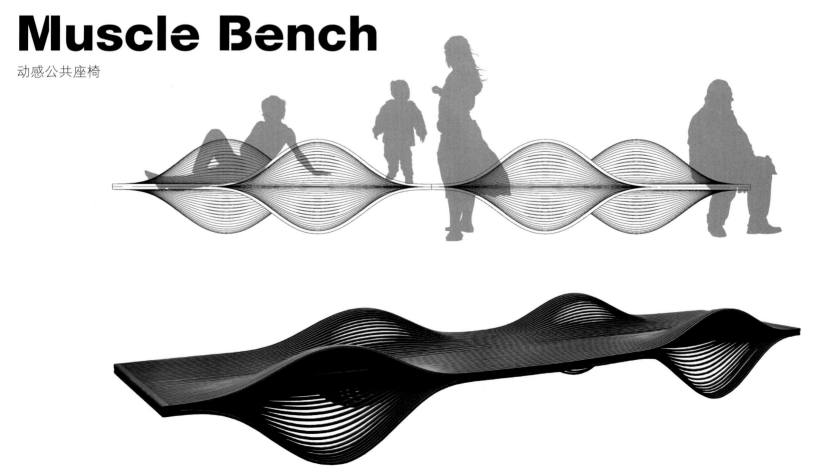

As spectacular to gaze at as it is comfortable to rest on, Muscle is a counterproposal to traditional stiff and motionless street furniture thanks to its dynamic forms and its purified lines. A piece to compliment both contemporary architectural structures or modern landscape designs. The bench offers the possibility of sitting or lying down in response to the surface's relief. Like the fibrous structure of a muscle, the cut metal sheets work with compression and tension to maintain the rigidity of the resting platform. The smooth gun metal grey epoxy finish reflects the light and adds to the sense of lightness. As the sun passes through the metal bars, the shadows mix with the actual piece adding yet another dimension to this exceptional urban bench.

既是一道引人驻足凝望的风景，又是一处舒适的休息场所，公共座椅 Muscle 采用动态的外形和简明的线条，突破了传统街头设施死板、冷酷的形象。它在当代建筑结构设计和现代景观设计两方面都获得了好评。

得益于表面缓解人体压力的设计，座椅给人们提供坐着或躺下的机会。金属条就像肌肉表面的纤维，通过伸力和张力作用固定在一起，确保了座椅平面的稳固。

光滑的青灰色环氧面漆反射着光线，使座椅增添了明亮的感觉。

当阳光穿过金属条投出的斑驳的倒影，连同金属条本身为这座独特的城市公共座椅增添了空间。

DESIGNER	DESIGN COMPANY	PHOTOGRAPHER
Damien Gires	Le Plan B - Architecture, Lumière, Intérieur, Design	Justin Westover

Urban Seat

都市座椅

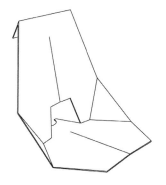

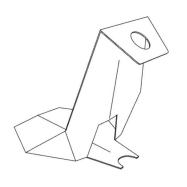

More than a concept, Urban Seat is a new opportunity to reclaim the urban pedestrian areas.

Urban Seats are lightweight, sturdy and stackable products designed to be placed on the anti-parking poles that line the streets. Urban Seat, naturally nomadic, finds its place in front of shops, galleries or offices.

This chair allows the creation of cafés ephemeral islets exchange and participate in the comfort of life for residents and pedestrians.

It is mounted in an instant. In the life of the city, the comfort of urban spaces helps to improve the quality of life and the emancipation of social relations.

Urban Seat is a creation of Damien GIRES. It is developed by the agency Le Plan B.

都市座椅不仅仅是一个概念，它是一个开拓都市步行区的新机遇。

都市座椅：它们重量轻，坚固，具有折叠性，这样的设计可以沿街摆放在禁止停车拦截柱上。

座椅自然地游荡着，在商铺、画廊、办公室的门前找到自己的位置。

座椅也可以折叠成咖啡桌来短暂使用，为市民和行人提供舒适生活。

它安装方便。在城市生活中，都市空间的舒适有助于提高生活质量，解放社会关系。

都市座椅是设计师 Damien GIRES 的创作。它由 LePlanB. 公司进行推广。

DESIGNER
Nichola Trudgen

PHOTOGRAPHER
Nichola Trudgen

Wanderest

Wanderest 座椅

The Wanderest is a seat or leaner that attaches to existing circular or octagonal lamp posts situated around rest homes and retirement villages. In order to keep the increasing elderly population healthy and mobile they need to do regular light exercise. Walking is an easy and cost-efficient way to keep active but the elderly often need to break their walk up into manageable distances with small intervals to rest. Lamp posts are an ideal resting point because they are at equal intervals along the footpath. If every lamp post along a street had a Wanderest it could be looked at as a goal for the elderly, to make it to the next resting spot.

The identical panels are injection moulded from a recycled Wood Plastic Composite. This sustainable material is durable, cheap and rot-resistant. The seat is installed at a perching height and attached either directly with a bolt or clamped on with a steel strap.

The Wanderest panels can be used on many different structures including flat surfaces. This means it could be used in areas where you have to wait for short periods of times such as banks and hospitals.

The design has won several prestigious awards such as an IDA Design Award, NZ Best Design Award and was a finalist in the International Dyson Awards.

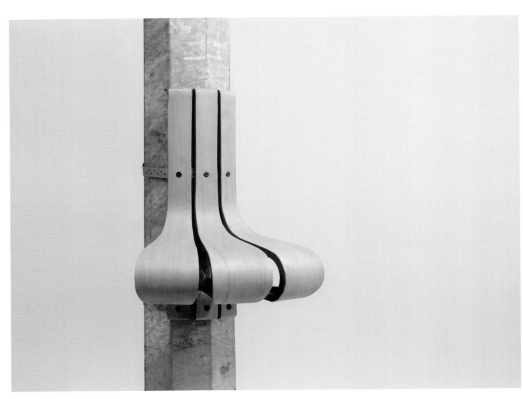

Wanderest 是一个固定在现有的圆形或八角形的灯柱上的座椅以及依靠点，它坐落在养老院和老年村。为了提高不断增加的老年人的身体健康和运动机能，老年人需要定期做轻微的运动。散步是一种保持精力充沛的简单且节省成本的有效方式，但老年人通常需要在路途中停留，将路程分为短暂间隔，在一定的距离内稍事休息。电灯柱是一种理想的休息点，因为它们是沿着人行道的间隔。如果在街上的每个灯柱有一个 Wanderest 座椅，老年人看作的一个目标，让他们坚持步行到下一个休息点。

相同的座椅面板从回收木材塑料复合材料注塑成型。这个可持续材料经久耐用，便宜且具有抗腐性。座椅安装在一个适合停靠的高度，用螺钉直接固定或用钢带夹在灯柱上固定。

Wanderest 面板可以被用在许多不同的建筑结构，包括平坦的墙面。这意味着它可以用在一些需要短暂等待的区域，如银行和医院。

设计赢得了多个奖项，如 IDA 设计奖、新西兰最佳设计奖，并是国际 Dyson 奖的入围作品。

ARTIST	DESIGN COMPANY	PROJECT DIRECTOR	CLIENT
Andrew Rowe, Simon Fenoulhet	D.A.R. Design	Mererid Velios, Celfwaith	Cardiff City Council

Delta Street

德尔塔街

Cardiff City Council identified Delta street as a new public space in the Canton area of the city that was in need of renewal. The space was created by closing a slip road and was newly landscaped to give more space back to the pedestrian. The brief was to design a seating feature that would provide a focal point in the space and encourage local people to use it fully, increasing the average "dwell" time. The artists designed a double seat that has an inside and outside edge so it can be used on either side. The design results in a social space inside the arcs for those who know one another and a more private outer arc for those wanting to sit on their own.

The seat consists of two C-shaped forms that interlock at the back and connect to the central foundation through a single foot. The seating platforms are cantilevered out from the centre, tapering in a shallow "V" to the ends. Fabricated from sheet steel, the structure forms a stable and rigid platform while appearing slender and lightweight. On the underside of the arcs is a series of blue LED lights that come on at dusk and cast an eerie light beneath the seat, emphasising its sculptural form at night and helping to enliven the space.

加的夫市议会确定德尔塔街是在坎顿地区一个新的公共场所，这地方需要整新。通过封闭高速公路枢纽整修，美化后的景观给予行人更广阔的空间。简要设计一个用于休息的座椅将会在这个地方引起关注，并鼓励当地人充分地利用它，提高日均使用的时间。艺术家设计了一张双面椅，有内外边缘，两侧都能坐人。设计的结果是：弧内的座位提供给那些彼此认识的人，他们需要的社交空间小；弧外的座位则提供给那些需要更多的私人空间的人。

座椅由两个"C"形组成，后面是联结在一起的，椅子的中间部分共用一个座脚。座位表面是从中间悬臂式地延伸出来的，以微"V"形逐渐收尾。用钢板焊接，这个结构形成了稳定而坚固的椅面，显得修长而轻盈。弧形的下方是一连串的蓝色 LED 灯，在黄昏时分，椅子下方就会放出奇异的光，在夜间突出其形状，活跃场地的氛围。

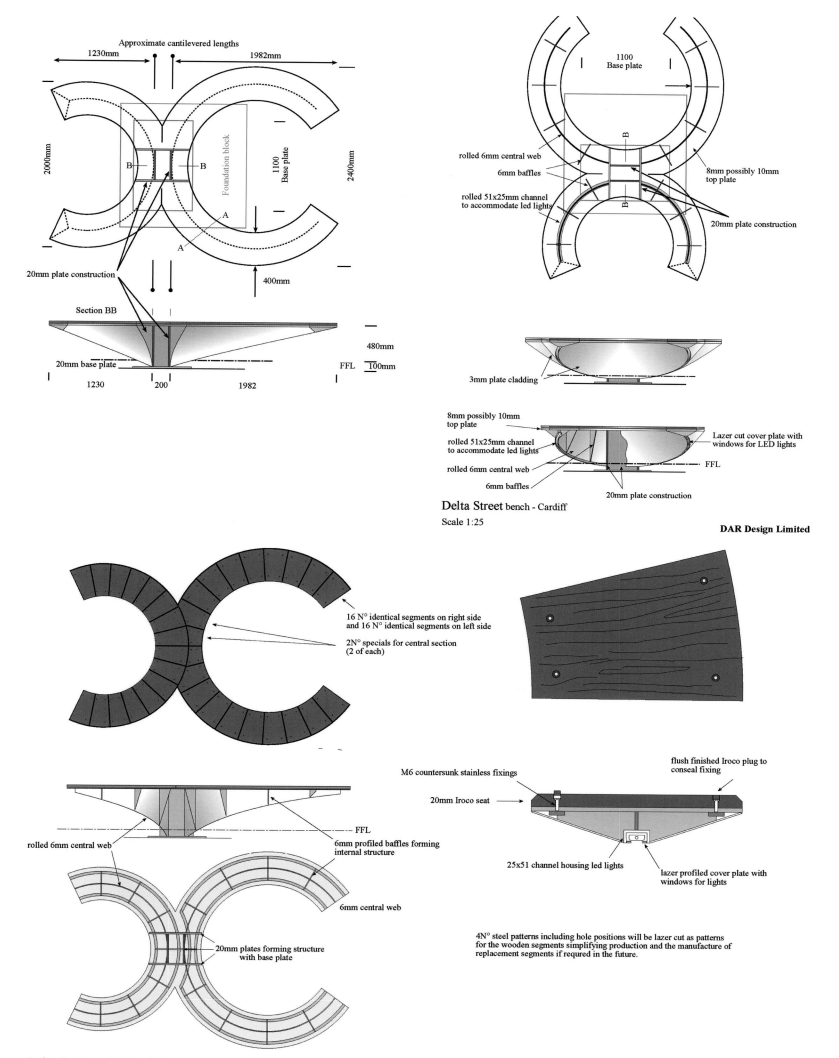

Approximate cantilevered lengths

1230mm   1982mm

2000mm

B — B

Foundation block

1100 Base plate

2400mm

A

A

20mm plate construction

400mm

Section BB

20mm base plate

1230   200   1982

480mm

FFL   100mm

1100 Base plate

rolled 6mm central web

6mm baffles

rolled 51x25mm channel to accommodate led lights

B

B

8mm possibly 10mm top plate

20mm plate construction

3mm plate cladding

8mm possibly 10mm top plate

rolled 51x25mm channel to accommodate led lights

rolled 6mm central web

6mm baffles

Lazer cut cover plate with windows for LED lights

FFL

20mm plate construction

Delta Street bench - Cardiff

Scale 1:25

DAR Design Limited

16 N° identical segments on right side and 16 N° identical segments on left side

2N° specials for central section (2 of each)

FFL

rolled 6mm central web

6mm profiled baffles forming internal structure

6mm central web

20mm plates forming structure with base plate

flush finished Iroco plug to conseal fixing

M6 countersunk stainless fixings

20mm Iroco seat

25x51 channel housing led lights

lazer profiled cover plate with windows for lights

4N° steel patterns including hole positions will be lazer cut as patterns for the wooden segments simplifying production and the manufacture of replacement segments if requred in the future.

Delta St Artwork, Cardiff

DAR Design Limited

DESIGNER	CLIENT	MANUFACTURER
Ignacio Ciocchini, IDSA. Industrial Designer	New York City Department of Transportation	Landscapeforms / Studio 431

CityBench

城市长椅

CityBench makes the sidewalks and streets more accommodating to senior citizens, the mobility-impaired and other pedestrians of New York City. Existing public seating systems seldom consider different body types, or allow for an acceptable distance or social space between multiple users. The CityBench provides comfortable seats with generous widths to accommodate a diverse population; a place to relax, read, eat and converse. To ensure adequate social space, the designer placed two dividers on each bench, creating three discrete personal seats, each with generous proportions of 67.95cm wide. Users with different body types can sit, change positions, and make other movements without bothering or touching their neighbour. Most can also place belongings next to them – purses, bags, bottles, lunch boxes, etc. CityBench features an asymmetrical laser-cut pattern that seems to be "moving" performing an important identifier role by relating the bench to New York City's energy and constant movement, and also to its skyline.

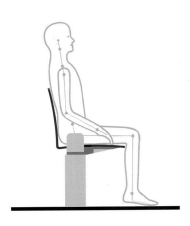
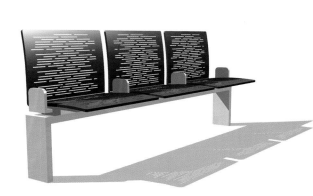
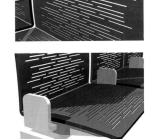

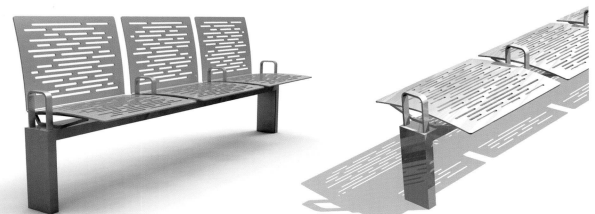

MATERIAL	DIMENSION	LOCATION	PHOTOGRAPHER
Laser-cut carbon steel plate and U channels E-coat and polyester powder coat finish	L 2.3m x W 0.6m x H 0.9m x Seat-H 0.5m	New York City, USA	Ignacio Ciocchini

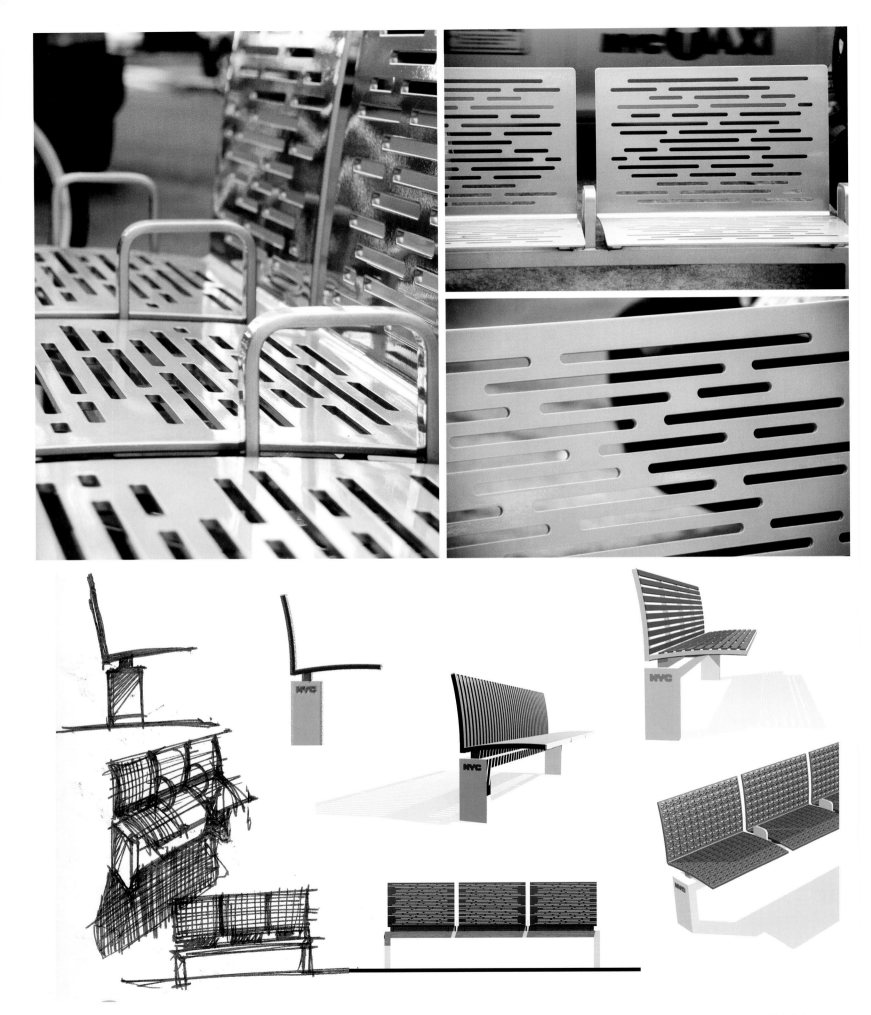

对纽约市的老年人、行动不便的人和其他行人来说，城市长椅让人行道和大街小巷更为人性化了。现有的公共座椅系统很少考虑体形各异的人，也很少允许多个使用者之间有个可接受的距离和社交空间。城市长椅提供了舒适的座椅和足够的宽度来适应不同的人群；一个放松、阅读、吃东西和交谈的地方。为了确保有足够的社交空间，设计师在每张长椅上做了两个分隔设置，做成了三个人的独立座位，每个座位有足够的比例——67.95 厘米宽。不同体型的使用者可以坐下，改变方向，在没有打扰和触碰旁人的前提下做其他动作。大多数椅子还有空间让使用者在他们旁边放置物品——钱包、袋子、瓶子、饭盒，等等。城市长椅设有一个不对称的激光切割图案，这似乎是"动态"上演着纽约市风貌和它持续的动感，同时也映衬了天际。

ARTIST	DESIGN COMPANY	CLIENT
Barbara Grygutis	Barbara Grygutis Sculpture LLC	MTA Art in Transit (Metropolitan Transit Authority), New York, New York

FABRICATOR	DIMENSION	MATERIAL	LOCATION	PHOTOGRAPHER
Art Metal Industries (AMI) Brookfield, CT	Five panels, 7'H x 4'W x 18"D each	Stainless Steel	The Bronx, New York	Peter Peirce

Bronx River View

布朗克斯河一瞥

Stainless steel sculptural seating and windscreens installed at the Whitlock Avenue Station on the Pelham Line, in New York City. Inspired by the location of the Whitlock Avenue Station, Bronx River View consists of five stainless steel sculptural units situated on the platforms and integrated into the windscreen walls. From floor to ceiling, the sculptures offer views to the Bronx neighbourhood and the nearby Concrete Plant Park and Bronx River. The functional sculptures blend the transit experience with the surrounding community and provide resting points with windows of open air and sky views. Bronx River View creates an environment to inspire dialogue based on the old and new and the traditional and contemporary. The New York Municipal Art Society awarded Bronx River View an Honourable Mention at the 2011 MASterworks Awards in New York City, New York.

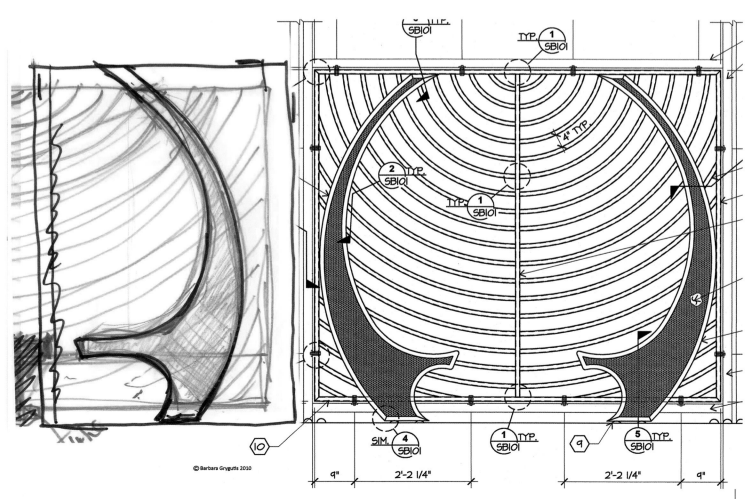

© Barbara Grygutis 2010

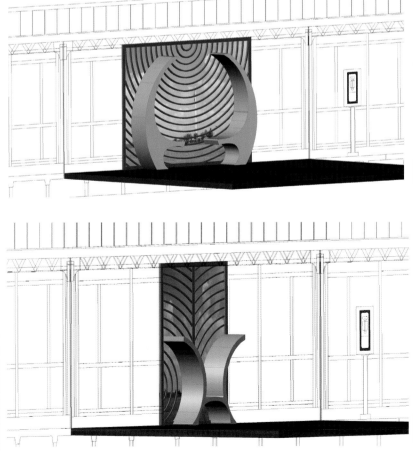

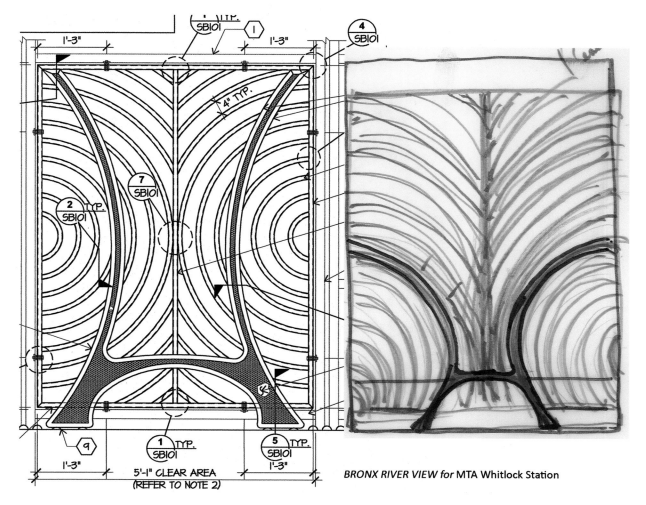

车站的位置给予了创作灵感，"布朗克斯河一瞥"景观包含了5个不锈钢雕塑部件，它位于站台上，与挡风墙融为一体。从地板到天花板，这个雕塑作品给人们提供了座位、布朗克斯河以及附近的混凝土植物公园的风景。这个功能性的雕塑景观将旅途辗转中人们的体验与周围的群落融合在一起，提供了户外新鲜空气与天空美景之窗的休息地。"布朗克斯河一瞥"景观营造了一个激发人们谈古论今、传统的与现代的聊天环境。在 2011 年的纽约市的 MASterworks 颁奖上，纽约市艺术协会授予"布朗克斯河一瞥"景观优秀奖。

BRONX RIVER VIEW for MTA Whitlock Station

DESIGNER
David Karásek, Radek Hegmon

DESIGN COMPANY
mmcité

Bistrot

餐桌椅景观

A completely new concept for public seating. Ideal for a quick meeting, a snack, or to send messages from your laptop, the increased seating area is truly comfortable, even when you're wearing a suit. Thanks to its compact size, it's suitable even for tight spaces. In combination with a table it can be used to create an interesting set-up for outdoor refreshment. With a rugged steel frame and a seating area or table top made from high-pressure laminate, it optionally can be decorated with a pattern or graphic design.

　　一个公共座椅的全新概念。开短会，吃零食，或者用便携式电脑发送信息的理想之所，即便你穿着西装，增大的座位面积真的非常舒适。由于其紧凑的尺寸，即使是在狭小的空间，也很适合。结合桌子，这个景观可以为户外茶点创新一个有趣的搭配。连接着坚固钢架的座位区或由高压板制成的桌面，都可以任意性地用图案或者图形设计来点缀。

DESIGNER
S Diamond, J Coughlan, K Wolak

DESIGN COMPANY
Stephen Diamond Associates

ENGINEER
Clifton Scannell Emerson Associates

LOCATION
University College Dublin, Ireland

Arts Block Entrance Deck

艺术街大门

The Arts Block Entrance Deck is a celebration of student life, conceived as a stage upon which students, staff and visitors interact.

Design inspiration was taken from the strict linear structure of the Arts Block elevations, which led to a linear framework of one-metre-wide strips of hardwood punctuated by sculptural benches. Carefully detailed blue acrylic end sections utilise the benches as a series of navigation beacons, arranged to draw attention to the new space and channel people towards the Arts Block entrance. This injection of colour acts as a counterpoint to the predominance of concrete on campus.

The benches establish a dialogue with the existing trees to place emphasis on their retention and take advantage of the dappled shade beneath.

As you approach along the main campus promenade, the transition from the dull lifeless thud of concrete underfoot to the resonance of timber signals a theatrical "treading the boards" entrance to the Arts Block.

At nightfall the benches come to life as a procession of glimmering blue light boxes. To complete the university colours a halo of green and yellow up lighters silhouette the delicate form and movement of bamboo, ornamental grass and mahonia against the adjacent buildings, covered walkway and bridge.

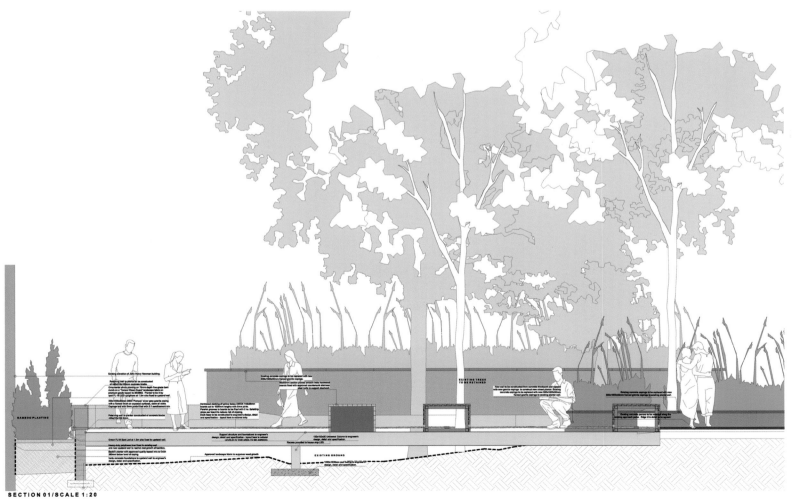

SECTION 01/SCALE 1:20

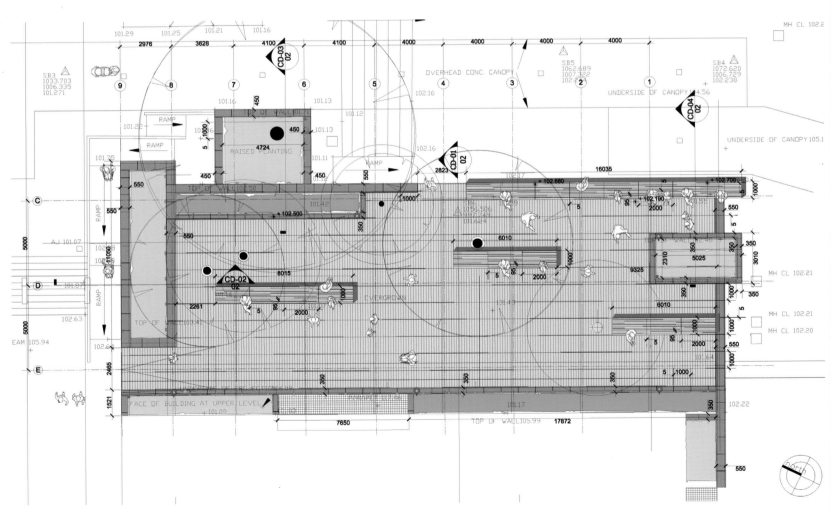

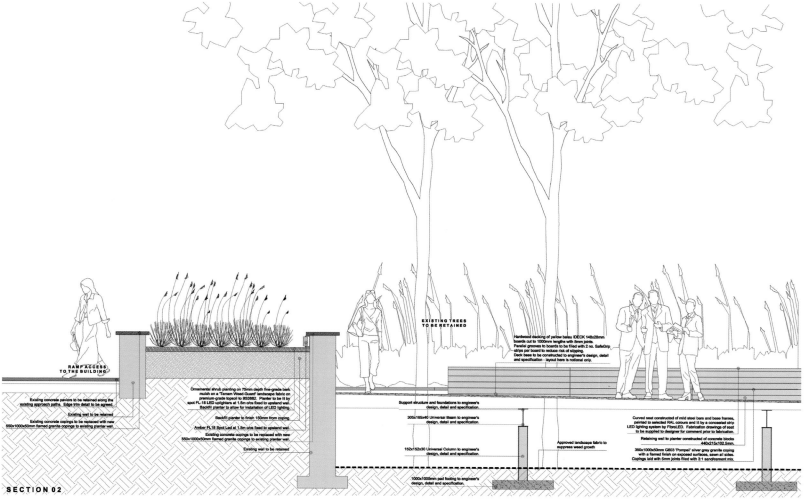

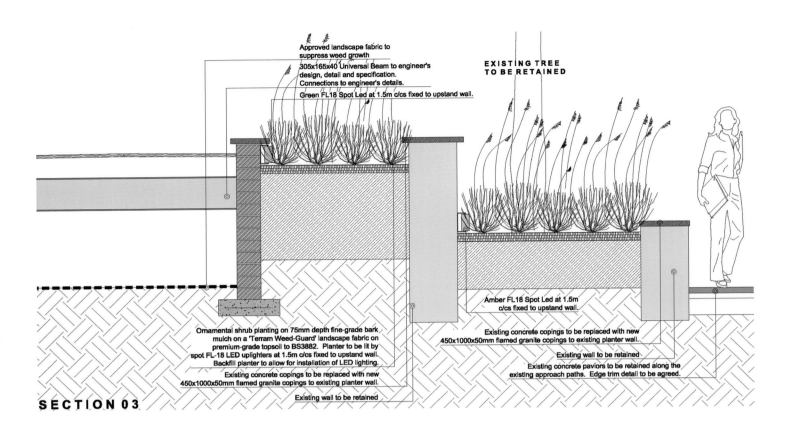

Approved landscape fabric to suppress weed growth

305x165x40 Universal Beam to engineer's design, detail and specification. Connections to engineer's details.

Green FL18 Spot Led at 1.5m c/cs fixed to upstand wall.

EXISTING TREE TO BE RETAINED

Amber FL18 Spot Led at 1.5m c/cs fixed to upstand wall.

Ornamental shrub planting on 75mm depth fine-grade bark mulch on a 'Terram Weed-Guard' landscape fabric on premium-grade topsoil to BS3882. Planter to be lit by spot FL-18 LED uplighters at 1.5m c/cs fixed to upstand wall. Backfill planter to allow for installation of LED lighting.

Existing concrete copings to be replaced with new 450x1000x50mm flamed granite copings to existing planter wall.

Existing concrete copings to be replaced with new 450x1000x50mm flamed granite copings to existing planter wall.

Existing wall to be retained

Existing concrete paviors to be retained along the existing approach paths. Edge trim detail to be agreed.

Existing wall to be retained

SECTION 03

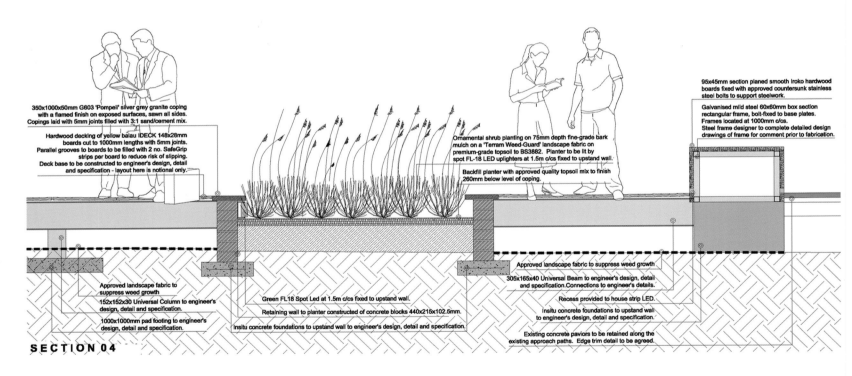

350x1000x50mm G603 'Pompeii' silver grey granite coping with a flamed finish on exposed surfaces, sawn all sides. Copings laid with 5mm joints filled with 3:1 sand/cement mix.

Hardwood decking of yellow balau IDECK 148x28mm boards cut to 1000mm lengths with 5mm joints. Parallel grooves to boards to be filled with 2 no. SafeGrip strips per board to reduce risk of slipping. Deck base to be constructed to engineer's design, detail and specification - layout here is notional only.

Ornamental shrub planting on 75mm depth fine-grade bark mulch on a 'Terram Weed-Guard' landscape fabric on premium-grade topsoil to BS3882. Planter to be lit by spot FL-18 LED uplighters at 1.5m c/cs fixed to upstand wall.

Backfill planter with approved quality topsoil mix to finish 260mm below level of coping.

95x45mm section planed smooth Iroko hardwood boards fixed with approved countersunk stainless steel bolts to support steelwork.

Galvanised mild steel 60x60mm box section rectangular frame, bolt-fixed to base plates. Frames located at 1000mm c/cs. Steel frame designer to complete detailed design drawings of frame for comment prior to fabrication.

Approved landscape fabric to suppress weed growth

152x152x30 Universal Column to engineer's design, detail and specification.

1000x1000mm pad footing to engineer's design, detail and specification.

Green FL18 Spot Led at 1.5m c/cs fixed to upstand wall.

Retaining wall to planter constructed of concrete blocks 440x215x102.5mm.

Insitu concrete foundations to upstand wall to engineer's design, detail and specification.

Approved landscape fabric to suppress weed growth

305x165x40 Universal Beam to engineer's design, detail and specification.Connections to engineer's details.

Recess provided to house strip LED.

Insitu concrete foundations to upstand wall to engineer's design, detail and specification.

Existing concrete paviors to be retained along the existing approach paths. Edge trim detail to be agreed.

SECTION 04

艺术街大门是学生生活的场所，是学生、员工、游客交流的平台。

设计灵感来自精密线性结构的艺术街的大门，引向由一条1米宽的硬木条和雕刻的长凳的线状构架。详细的蓝色丙烯酸端部利用长凳作为一系列的导航指标，意在引起人们对艺术街大门的新场地和通道的注意。这种艺术色彩的注入，是校园混凝土建筑优势的具体对比。

长椅与现有的树木间建立了对话，仿佛要重点强调它们的记忆，并利用树荫下的斑驳阴影。

当你走在校园的主长廊，从脚下毫无生气的混凝土到与脚产生共鸣的木板的过渡，标志着戏剧性地"踩在木板上"进入艺术街。

夜幕降临，长椅仿佛一排排发出微蓝光亮的盒子，充满了生机。为了完善校园而用投射灯涂上了晕绿色和黄色，显现了微妙的形式。摇摆的竹子、观赏性的草和丛生的植物，映衬了附近的建筑，笼盖了通道和天桥。

UrbanEdge

UrbanEdge 城市边缘

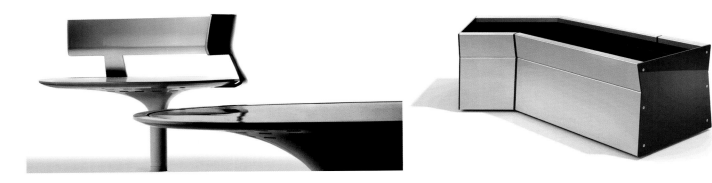

Seattle, WA (June 1, 2012) - Gustafson Guthrie Nichol (GGN) is pleased to announce the introduction of UrbanEdge, a collection of landscape framing and furniture elements for urban spaces.

The furnishings respond to the demand for more outdoor public spaces with an interest in less formal, more welcoming and effective use of scarce available urban locations. UrbanEdge is a tool for designers to create unique yet functional sites; defining edges and transforming inactive corners into outward-oriented settings that interface with the urban fabric. It constructs places for orienting and organising, resting and reflecting, meeting and greeting, people watching and taking in the view. It can be used to create curbside transit stops and "eddies" along sidewalks where people step in and out of the flow; to develop niches and focused social settings within larger spaces; and to enrich experience in outdoor spaces with dynamic and kinetic furniture elements that suit the way we interact today.

The UrbanEdge family members are the MAX Trellis; JESSIE Rail; GUS Planter; OLLIE Small Seat; SOPHIE Large Seat; BERNIE Bar Height Seat; and STELLA Table.

Frames include the trellis, rail system and planter. The core elements are the seating and table pieces that live within defined spaces and activate the centre, enhancing the experience of place.

UrbanEdge's elements are multifaceted and work together in various combinations. The suite of furnishings was created by designers for designers, and focuses on the functionality of spaces and the facilitation of a wide range of activities. The different pieces can be combined in inventive ways to activate outdoor settings and provide multiple opportunities for people to interact with in the same space over time. UrbanEdge is a dynamic resource for maximising under-leveraged sites, designing vibrant, integrated urban settings and creating a sense of place.

2012 年 6 月 1 日在华盛顿州西雅图市，古斯塔夫森·格思里·尼科尔设计公司（GGN）欣然向公众推出"城市边缘"，即都市空间中的景观框架和设施系列产品。

这些景观设施使有限的城市空间变得不再生硬拘谨，而是更加热情高效，以此解决了人们对户外公共空间的需要。"城市边缘"使设计师创造出独特而又具有功能性的场所；划清边界并将闲置的角落转变成外向型场地，与城市结构相融合。人们可以在这里或看清方向或组织活动，或安逸休息或沉思默虑，或见面会晤或寒暄问候，或驻足观望或欣赏美景。它可以用作路边交通站，也可以是人行道边的"旋涡"式景点，人们从人流中出来在这里小憩；在较大的空间里形成小环境和集中的社交场合；运用动态活跃的、与现在生活方式相适应的装置元素丰富户外空间体验。

"城市边缘"的家族成员有马克思格子结构、杰西横杆、格斯花架、奥利小型座椅、苏菲大型座椅、伯尼酒吧高度座椅和斯特拉桌子。

框架包括格子结构、横杆架系统和花架。核心元素是座椅和桌子，它们在有限的空间内赋予该休闲区域以生命，加强了空间体验。

"城市边缘"的元素是多层面的，在不同的组合中相互作用。一整套设施都是出自设计师之手，注重空间的功能性，有利于广泛的活动。不同的板块可以以一种别出心裁的方式相互结合，使户外设施具有活力，即使人们在不同时间到来，也都会与这同一空间产生不同的互动体验。"城市边缘"是一种动态资源、优化较低层次的景点，设计出充满活力、和谐的城市环境，创造出一种地区景观感。

DESIGNER
Alexandre Moronnoz

COLLABORATOR
Chartier-Corbasson Architectes

LOCATION
Lille, France

Snake Seating

蛇形座椅

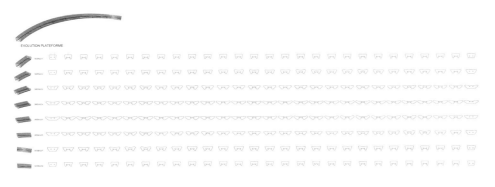

Furniture for multiple functions

The sculptural aspect of the project reflects the decorative elements of the surrounding façades.

In order to offer the possibility of multiple uses, the snake project is designed like a long ergonomic morphing, evolving for example from bench to platform. The successive assembly of different profiles re-forming the formal evolution of the seat creates a kinetic effect which peps up the large pedestrian crossings on the square. The three large snake project semicircles give the possibility of reconstructing an architectural pattern on the scale of the square, over the years, by moving them. They are accompanied by additional fixed parts designed around the pedestrian entrances to the underground car park. These parts offer, amongst other functional additions, a space designed for bicycle storage.

Cast concrete standard base and metallic inserts

Metallic feet, planimetric adjustment

Pyramidal pedestal, water and residue evacuation

Variable wood and metal superstructure

Modular assembly

Thin fold CP laser-cut frames

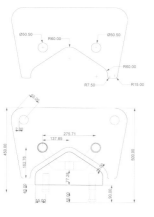

PROFIL BASE

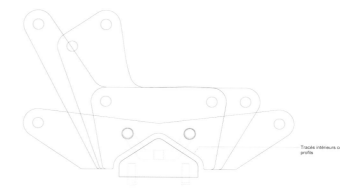

Tracés intérieurs communs à tous les profils

PROFILS TYPES

多功能家具

设计富有雕塑感的造型，对周围环境起了装饰作用。

为了创造多功能座椅的可能性，蛇形座椅设计成类似一个符合环境改造学的长形的变体结构，例如从座椅到平台演变。不同大小纵向剖面连续的排列，重新组合，完成了座椅形态上的变化，创造出一种动力学的效果，形成了在广场上集中的步行区域。通过移动三个大的蛇形围绕半圆，多年来，给广场规模的建筑结构的重新塑造创造了无数可能性。同时也伴随着对步行入口的地下停车场的其他固件的使用。除了其他功能的增加，这些部件还提供用来存放自行车的空间。

混凝土铸造规格基层和金属填充物

金属底座，平面调整

金字塔的基座，水渣疏散

可变的木材和金属结构

模块化组件

薄薄的 CP 激光切割框架

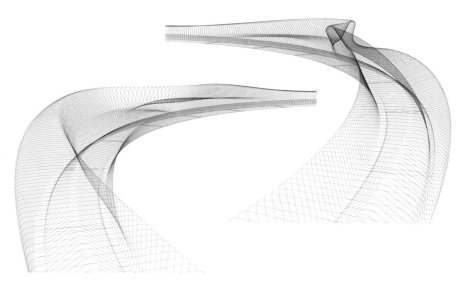

DESIGNER	FIRM	CLIENT	LOCATION	PHOTOGRAPHER
Gerhard Nijenhuis, Martijn van Dongen	ipv Delft	Falco.nl	Leeuwarden, the Netherlands	Henk Snaterse

XXL Street Sofas

XXL 街头沙发

Sundeck, sofa, picnic table or stage. These XXL street sofas designed for a large square in Leeuwarden's city center, combine different functions. With their three different seating levels (43, 58 and 65 centimeter respectively) each user can choose his own manor of use.

Ipv Delft designed the sofas on the basis of a sketch by Hosper landscape architecture and urban design, which made the overall design for the refurbishment of Wilhelmina Square. Each 5 by 2 metre sofa is placed around a tree. There are two versions of the sofa: one where the tree is placed in the center and one where it isn't. Detailing and materials are identical though. The tree is placed inside a circular opening which is closed off by black plastic sheeting that follows the outline of the tree. At night, integrated lighting within a steel ring around the tree illuminates the trunk, branches and leaves.

The XXL street sofa is made out of wood and steel, a timeless combination with a high-end appearance, which makes it a perfect example of ipv Delft design.

日光浴场所、沙发、餐桌或休闲台，这些 XXL 街头沙发是结合了不同的功能，为在莱瓦顿市中心的大广场设计。它有三个不同高度的座位（分别为 43 厘米、58 厘米和 65 厘米），每个用户可以选择自己的位置。

设计师 IPV Delft 在 hosper 景观建筑和城市规划事务所提供的概况的基础上设计了沙发，为 Wilhelmina 广场翻新做了全面设计。每隔 2 米就有 5 个沙发被放置在树的周围。有两种方式：一种是树木放置在沙发中央；另一种是中央没有放置树木。两种方式摆放的座椅细节设计和选用的材料都是相同的。树上放置一个圆形开口，用黑色塑料片沿着树木轮廓封闭。在晚上，围绕树木的钢圈里的集成灯光照亮了树木的树干和枝叶。

XXL 街头沙发采用木料和钢铁制成，是一个拥有一种高档外观的永恒的座椅组合，它成为设计师 IPV Delft 的一个杰出设计。

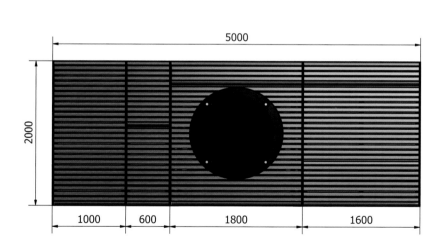

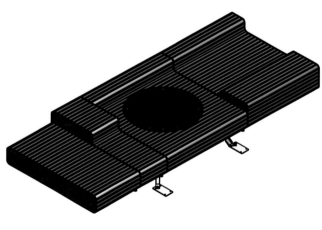

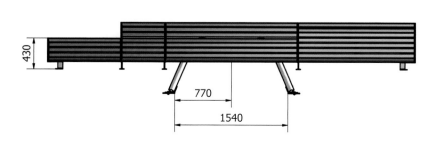

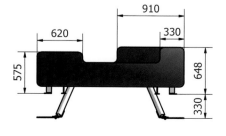

DESIGNER
Diana Cabeza

PRODUCTION AND COMMERCIALIZATION
Estudio Cabeza

DEVELOPMENT TEAM
Diana Cabeza, Martín Wolfson, Juan José Cano, Diego Jarczac

Topografico Bench

Topografico 公共座椅

Sinuous, unsymmetrical and irregular, this bench's undulating surface slides like the earth, glides like wet sand and draws watermarks as on a still damp surface.

Its topographic expression evokes subtle ergonomic qualities. Its concrete materiality is user-friendly. The bench can be configured in numerous and varied ways.

It is available with or without backrest, in straight or curve version.

Materials

Precast concrete with colour aggregate. Natural finishing.

Curve version can be acquired in two different radius. It can be matched side-to-side in the same radius or in combinations of both. It is currently available in black and brown grey.

Fixing

It can be simply placed or fixed with chemical or mechanical anchors.

Award: Good Design Label granted by the National Design Plan of the Argentine Ministry of Industry on 2012

DIMENSION

L 1.80m × D 0.70m × H 0.40m

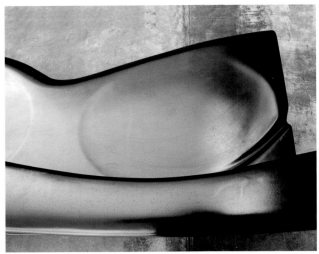

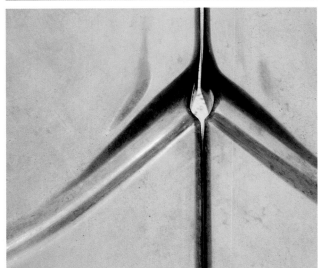

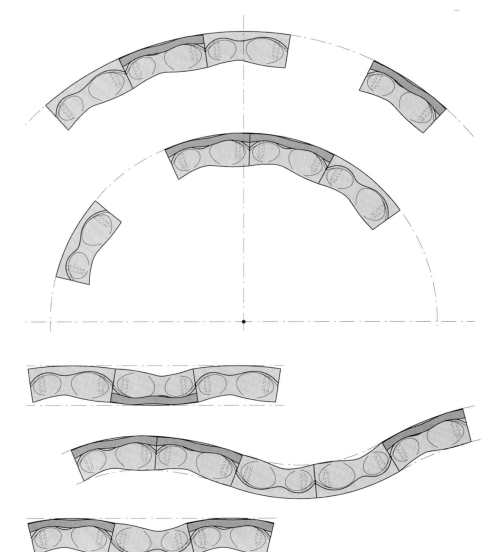

Single bench
without backrest

Single bench
with backrest

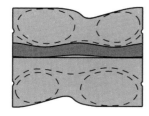
Area Arrangement
of benches
with and without
backrest

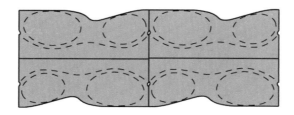
Row Area Arrangement of benches
without backrest

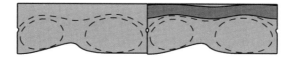
Row Arrangement of benches with and without
backrest

弯曲、非对称性、不规则，这把座椅的波状表面像球体一样平滑，材质就像潮湿的沙子那样细腻。

其地表外形表达了微妙的符合人体工程学的特性，它的混凝土材质深受人们喜爱。它可以有许多不同的方式和数量组合。

不论是否装有靠背，直线或是曲线排列，Topografico 公共座椅都可以被使用。

材料：

预铸混凝土材料形成座椅的颜色。天然成型。

曲线版可以在两个不同半径获得，它可以由相同半径的邻近的两个座椅或两个座椅叠加起来形成。

目前有黑色和棕灰色两种颜色。

固定：它可以简单地放置或用化学或机械锚固定。

奖项：在 2012 年阿根廷工业部国家设计计划中被授予杰出设计称号。

DESIGNER
David Karásek, Radek Hegmon

DESIGN COMPANY
mmcité

Radium

Radium 系列座椅

This line derives its benefit from the aesthetics of bent steel sheet. The high rigidity and overall excellent resistence of this bench is achieved thanks to an ingenious intersection of walls. The newly designed extended version features a more slender elegance and is available in all variations of materials. The basic all-steel style with dot perforation characterises the best in the Radium design. The combination of the unique "Citepin" seat pattern and plastic "button pads" thermally isolates the unpleasantly cold metal while enhancing the comfort of the seat. Solid wooden boards on the seat still constitute officially recognised and certified classics. The high pressure laminated HPL scores with an elegant smoothness, broadening the bench's interior installation possibilities. This product line also includes litter bin and bollards.

　　线条汲取了薄钢条弯曲弧度的美感。座椅的结实构造以及整体优良的抗压性得益于与墙壁接口处的巧妙设计。最新设计的改良作品略添雅致，采用多种种类不同材料建造。最初全钢制作配以打孔装饰的设计堪称 Radium 系列座椅中的经典款式。结合独特的"Citepin"座位图案以及塑料"按钮坐垫"，将金属质地那令人不愉快的冰冷感一扫而光，提升了座椅的舒适度。实木板款式上仍然是官方赞赏和认证的经典。高压积层板 HPL 呈现出优雅的光滑性，拓宽了座椅内部构造的多样性。该公司设计的产品还包括垃圾箱以及矮柱。

DESIGNER	DESIGN COMPANY	COLLABORATE	PHOTOGRAPHER
Pepe Gascón	Pepe Gascón Arquitectura	ESCOFET	Eugeni Pons

Equal Bench

Equal 长椅

Wooden and concrete bench

Cast stone and wood are combined in EQUAL, an element with a linear geometry and ample space for sitting and sharing that emerges from its proportion and balance. Balance in its volumes, which form a suite that shapes the base and emerges as a backrest. Balance between its inclined and orthogonal planes, creating a simple, surprising geometry. Balance between two materials that merge to heighten their contrast: the coldness of cast stone and the warmth of wood. Finally, balance between its potential use as a seat with and without a backrest. The nearly 3-metre length of this product provides a comfortable, attractive surface for sitting, while its weight allows for unanchored installation.

Materials and specifications:

Pickling and waterproofed concrete protected with lasur. Bolondo wooden seat

(1) Reinforced cast stone, standard colour chart, acid-etched and waterproofed

(2) Natural Bolondo wood (FSC)

Free-standing / 1.140 Kg

SISTEMA DE COLOCACION

elevación / hoisting
(P=1140 kg)

eslingas de poliester
polyester slings

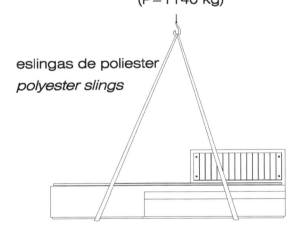

BANCO APOYADO SIN ANCLAJE
FREE-STANDING ELEMENT

木板和混凝土长椅

　　铸石和木材相结合做成了 Equal 长椅，线性几何与宽敞空间的要素提供了坐席和分享的空间，突出了长椅的比例与平衡感。体积的平衡，构成了一套由基石和浮木形成靠背的组合。斜度和垂直平面之间的平衡，造就了一个简单的、令人惊讶的几何形状。冰冷的铸石和温暖的木材这两种材料之间的平衡合并到一起，提高了它们的对比度。最终，它潜在的平衡作为一个座位使用，有靠背的和没有靠背的。接近 3 米长的长椅提供了一个舒适的、吸引人的坐席，而它的重量允许非固定性安置。

材料和规格：

耐酸和防水的混凝土用 lasur 保护的博隆多木质椅

（1）钢筋铸石板、标准彩色图表、抗酸并防水的

（2）天然博隆多木材（FSC）

高度不固定 /1.140 公斤

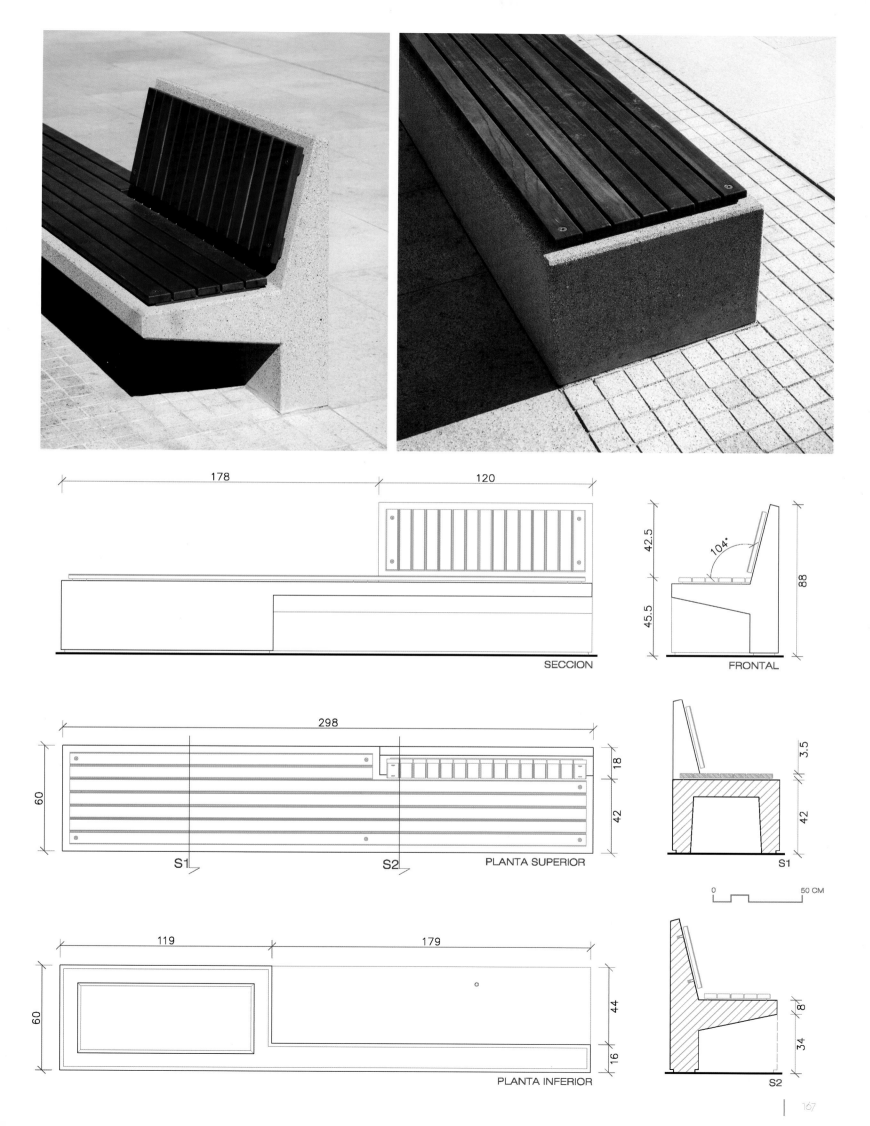

178   120

42.5   104°   88

45.5

SECCION   FRONTAL

298

60   18   42

S1   S2   PLANTA SUPERIOR

3.5

42

S1

0   50 CM

119   179

60   44   16

PLANTA INFERIOR

8

34

S2

DESIGNER	DESIGN COMPANY	MANUFACTURER	MATERIAL
Alexander Lotersztain	Studio Derlot	Street and Garden Furniture Company	mild-steel

LOCATION	PHOTOGRAPHER
Gold Coast, Australia	Florian Groehn

Marine Series

航海系列

The Marine Series is Brisbane-based designer Alexander Lotersztain's latest venture into the field of street and parkland furniture.

The designs take inspiration from the aesthetics and craftsmanship of luxury yachts, combining sleek lines and cutting-edge materials to form comfortable seating arrangements and refuse bins that offer the versatility of modular design.

The range of benches have been designed to cantilever like the bow of a sailing boat, complimenting the vessels moored in Sanctuary Cove's adjacent Marina in the Gold Coast, Australia.

The benches are made from mild steel that has been hot-dip galvanised and powder-coated in Lemon Yellow. The back and armrests are thermoformed DuPont Corian finished in Snow White, selected for its UV stability and corrosion resistance. 316-grade stainless steel has been used for the garbage and recycling bins as well as the seat fittings in order to ensure resistance to stains and corrosion.

The goal was not only to compliment the local surroundings, but also to enhance the visual identity of the marina and town centre developments through the design of a series of street furniture which responds directly to the context and lifestyle within a seaside community.

"航海系列" 是布里斯班设计师 Alexander Lotersztain 在街道和公共场所设备领域上的新进军。

设计灵感来自豪华游艇的美学和工艺，结合圆润的线条和刀口材料，造成了舒适的排座和垃圾箱，为模块设计提供了多样化。

长椅的椅面被设计成悬臂式，就像帆船的船头，称赞着停泊在澳大利亚黄金海岸神仙湾附近码头的船只。

长椅是由镀锌和柠檬黄的粉末涂层的低碳钢做成的。靠背和扶手加热塑形为杜邦可丽耐的雪白色，选择了 UV 稳定性和耐腐蚀性。316 级不锈钢已用于垃圾回收箱以及座椅配件，以确保抗污渍和抗腐蚀。

设计师的目标是不仅要称赞当地的环境，还要提高码头的视觉感，镇中心通过一系列街道设备的设计，直观地反映了海域边社区的环境和生活方式的发展。

DESIGNER	MANUFACTURER	LOCATION
Ignacio Ciocchini, IDSA. Industrial Designer	Landscapeforms / Studio 431	Chelsea, New York , USA

Chelsea Bench

切尔西长椅

问题：在附近的人行道上没有可坐的地方。纽约街区在使用中的长椅设计，没有给路人提供足够的空间让其感觉舒适；路人认为他们与旁人坐得太近了。

解决方案：新式长椅的优势

• 宽大的座位给路人提供了大量的"社交空间"

• 路人可以在不打扰旁人的前提下改变座椅位置

• 路人可以在自己的座位上放置物品，而不需占用另一个位置来放置一个袋子或者一个电脑

• 分隔的独立座位增加了私人空间而又不阻碍社交

该长椅的设计是如此的成功，以至于被纽约市选中，并延用至全市范围，作为纽约市交通部发起的"城市长椅方案"的一部分。

Problem: There were no sitting areas in the sidewalks of the neighbourhood. Bench designs in use in other New York neighbourhoods did not provide enough space for users to feel comfortable; users felt they were sitting too close to each other.

Solution: Advantages of the new benches

• The seats are very wide providing plenty of "social space" between users.

• Users can change seating positions without risking bothering a neighbour.

• Users can place objects on their own seat without the need to block an extra seat to locate a bag or a computer.

• Separated individual seats reinforce the idea of personal space without impeding social interaction.

The design of this bench was so successful that it was chosen by the City of New York to be extended citywide as part of the CityBench Programme launched by the NYC Department of Transportation.

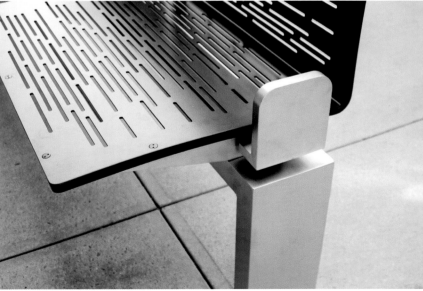

MATERIAL	DIMENSIONS	PHOTOGRAPHER
Laser-cut powder-coated carbon steel plates and bars, electro-polished and bead-blasted stainless steel U channels and plates	L 2.3m x W 0.6m x H 0.9m x Seat-H 0.5m	Ignacio Ciocchini

DESIGNER	DESIGN COMPANY	DIMENSION	PHOTOGRAPHER
Igor Solovyov	Simplicity Urban Solutions	2,230 × 352 × 380	Simplicity Urban Solutions Company

Melbourne

墨尔本长椅

The inspiration for Melbourne bench was megalopolis with straight avenues, parallel lines, and the buildings. The base of the bench is simple but yet elegant. Clear silhouette is ideal for urban landscape. The bench is made of cast aluminium and pine or ash wood in four colours. This project is designed by Simplicity Urban Solutions Company in 2010.

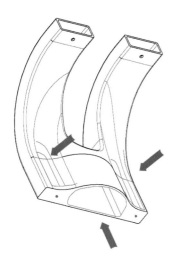

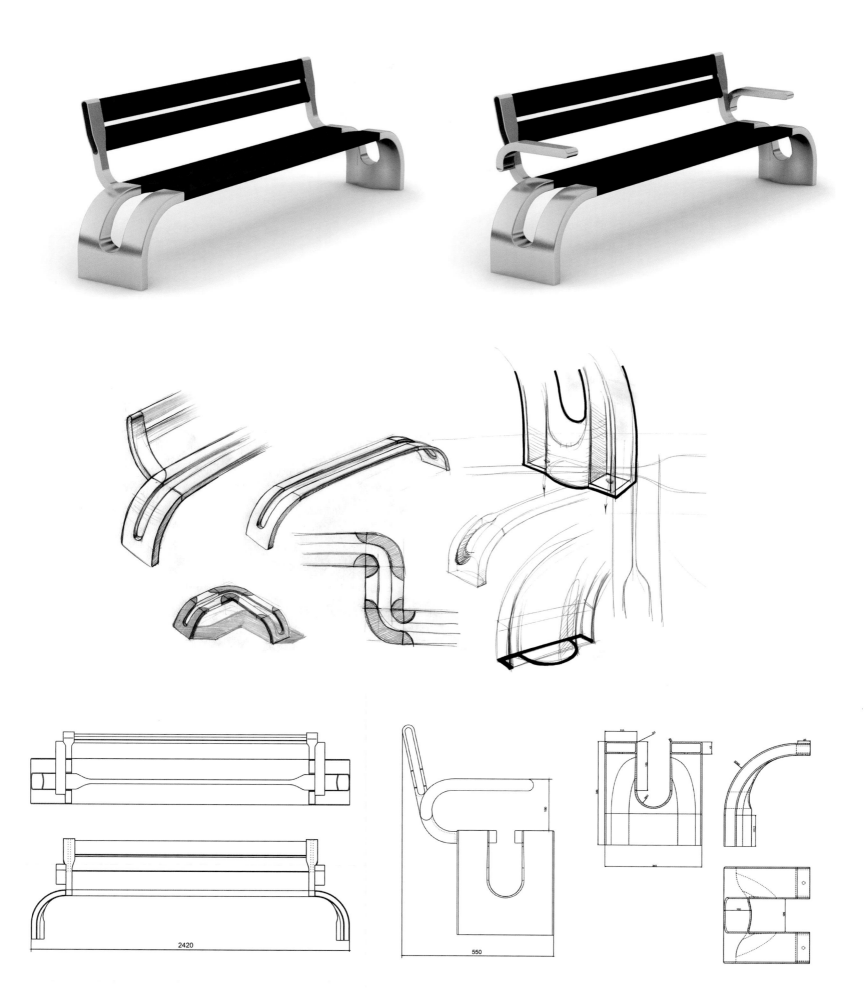

墨尔本长椅的设计灵感来自笔直的大道、平行线、大都市的建筑。长椅的底座简单而不失华丽。清晰的轮廓是城市景观的典范。长椅是采用铸造铝和 4 种颜色的松木或白蜡木制成的。这个项目于 2010 年由 Simplicity Urban Solutions Company 公司设计。

Patrimonial Bench

世袭的长凳

This piece of urban furniture was designed especially for the historical centre of the City of Buenos Aires.

It is made of precast concrete with an off-white incorporated colour aggregate, typical of turn-of-the-century Buenos Aires façades.

With or without backrest and the base unit measuring 1.40 metre long, it is possible to install individual units or assemble a long continuous row of benches. Furthermore, as benches with or without backrests can also be combined for approach from either side, very singular configurations may be obtained.

DESIGNER	PRODUCTION AND COMMERCIALIZATION	DEVELOPMENT TEAM
Diana Cabeza	Estudio Cabeza	Diana Cabeza, Alejandro Venturotti and Diego Jarczak

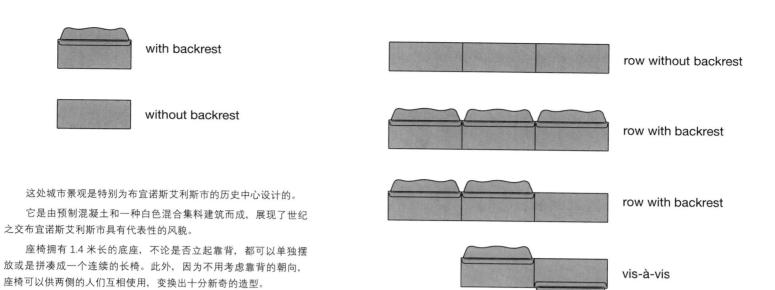

with backrest

without backrest

row without backrest

row with backrest

row with backrest

vis-à-vis

这处城市景观是特别为布宜诺斯艾利斯市的历史中心设计的。

它是由预制混凝土和一种白色混合集料建筑而成，展现了世纪之交布宜诺斯艾利斯市具有代表性的风貌。

座椅拥有 1.4 米长的底座，不论是否立起靠背，都可以单独摆放或是拼凑成一个连续的长椅。此外，因为不用考虑靠背的朝向，座椅可以供两侧的人们互相使用，变换出十分新奇的造型。

DESIGNER
David Karásek, Radek Hegmon

DESIGN COMPANY
mmcité

Portiqoa

Portiqoa 公共座椅

The perfect bench for those who love smart shapes. Its softly curved sides are made of aluminium alloy which work together with diagonally constructed walls to create an exciting optical illusion. The bench touches the base with only a narrow edge and from some angles reveals some impossibly thin lines. The continuous lamella frame arches to below the seat so as to significantly strengthen the entire structure. Individually shaped armrests placed in the middle prevent anyone from lying down, just as the pair of armrests on either side make it easy to get up.

对于那些喜爱灵巧外观的人来说，Portiqoa 公共座椅是完美的杰作。它流畅的带有弧度的底座是用铝合金制作而成的，连同对角线朝向设计的靠背，给人们一种激动人心的视觉感官。椅面只与底座狭窄的边缘接触，在底座一些角度边框上，搭建着令人难以想象的窄木条。连续的薄木片框架在座椅下方成拱形，因此大大稳固了整个构造。独特形状的扶手设置在中间，防止有人直接躺在椅子上，也有设置在椅子两侧的成对扶手让人们可以轻松站起。

ARTIST	MATERIAL	PHOTOGRAPHER
David Brooks	Douglas fir, silver birch and hardware	Barney Hindle, Courtesy of Cass Sculpture Foundation

Picnic Grove

树林餐桌景观

Through his immersive sculptural installations David Brooks explores different perceptions and interactions with nature.

As part of the Cass Sculpture Foundation's new Fields Programme, Brooks has created "Picnic Grove", a sprawling work built out of custom-made outdoor wooden furniture and spread over the entire 18,000m of the Deer Hut Field. The dozens of picnic tables and garden chairs are constructed in an interlocking manner, with trees heedlessly growing through the furniture like opportunistic weeds. As the picnic tables traverse the field and impose themselves on the landscape, the trees perforate the structures like a verdant grove, creating ambiguity as to which is dominant.

While visitors are encouraged to utilise the installation for communal enjoyment, they will also find themselves negotiating the playful interruptions created by the erratic placement of the trees, fostering a similar sense of ambiguity as to who is imposing on whom.

　　尽管设施拥有雕塑般的外形，设计师 David Brooks 仍然探索着各种与自然融合、交流的方式。

　　作为卡斯雕塑基金会新领域计划的一部分，设计师 Brooks 设计了"树林餐桌"。它采用定制木质户外家具，不规则地伸展摆放，景观蔓延至 the Deer Hut Field 地区 18,000 米的整个区域。许多餐桌和园椅以连锁式构造，树木随意地穿过桌椅，就像不经意的杂草。当餐桌遍布草地，呈现出休闲聚会氛围的风景，树木也穿插进来，形成一个郁郁葱葱的天然风貌，这一切不禁使人迷惑：哪一个才是占主导的风景？

　　提倡游客共享桌椅设施，同时会发现他们要与在桌椅间顽皮阻碍、随意伸出的树木协商位置，这更加强了人们的疑惑：到底是谁妨碍了谁？

REST PAVILION
休息亭

DESIGNER
Jürgen Mayer H., Andre Santer, Marta Ramírez Iglesias

DESIGN COMPANY
J. MAYER H. Architects

CLIENT
Ayuntamiento de Sevilla und SACYR

Metropol Parasol

大都市阳伞

Permanent Collection of Museum of Modern Art, NY and Staatliche Museen zu Berlin, Preussischer Kulturbesitz, Berlin Permanent Collection of DAM, Deutsches Architekturmuseum Frankfurt, Germany Holcim Award, 2005, Winner Europe Bronze for Sustainable Construction Mies van der Rohe Award 2013, Finalist

"Metropol Parasol", the Redevelopment of the Plaza de la Encarnacíon in Seville, designed by J. MAYER H., became already the new icon for Seville, a place of identification and to articulate Seville's role as one of the world´s most fascinating cultural destinations. "Metropol Parasol" explores the potential of the Plaza de la Encarnacion to become the new contemporary urban centre. Its role as a unique urban space within the dense fabric of the medieval inner city of Seville allows for a great variety of activities such as memory, leisure and commerce. A highly developed infrastructure helps to activate the square, making it an attractive destination for tourists and locals alike.

The "Metropol Parasol" scheme with its impressive timber structures offers an archaeological museum, a farmers market, an elevated plaza, multiple bars and restaurants underneath and inside the parasols, as well as a panorama terrace on the very top of the parasols. Realised as one of the largest and most innovative bonded timber-constructions with a polyurethane coating, the parasols grow out of the archaeological excavation site into a contemporary landmark, defining a unique relationship between the historical and the contemporary city. "Metropol Parasols" mixed-use character initiates a dynamic development for culture and commerce in the heart of Seville and beyond.

PLEXI-MODEL
Werk 5, Photographer: Uwe Walter

TIMBER-MODEL
Finnforest, Aichach

STRUCTURE
Concrete, timber and steel

PRINCIPAL EXTERIOR
timber and granit

PRINCIPAL INTERIOR MATERIAL
Concrete, granit and steel

PHOTOGRAPHER
Nikkol Rot

纽约现代艺术博物馆与柏林国家博物馆永久收藏，普鲁士文化遗产，法兰克福德国建筑博物馆永久收藏，2005 年荣获德国 Holcim 建筑设计大奖，2013 年获得密斯·凡·德罗欧洲当代建筑奖提名。

"大都市阳伞"是对塞维利亚的德拉恩卡纳西翁广场的重新改造，该项目由 J.MAYER.H 设计，已经成为塞维利亚的新地标——展现了塞维利亚作为世界上最具吸引力的文化圣地之一的地位。"大都市阳伞"充分挖掘了德拉恩卡纳西翁广场的潜能，将其变成当代城市新中心。在塞维利亚市中心密密麻麻的中世纪建筑中，"大都

市阳伞"显得独一无二，这一设计考虑到了纪念、休闲与商业等众多活动的需求。高度发达的基础设施为广场增添了活力，使其成为游客与当地人的向往之地。

"大都市阳伞"方案采用令人印象深刻的木结构，伞下与内部设有考古博物馆、农贸市场、高架广场、多功能酒吧与餐厅。顶层则是全景露台。阳伞采用聚氨酯涂料，是工程规模最浩大、最具创新性的黏合木结构建筑之一。这里从当初的考古挖掘地一跃成为当代地标，表明了历史与现代城市之间的独特关系。"大都市阳伞"用途广泛，为塞维利亚中心及周边地区文化与商业的发展注入了一丝新的活力。

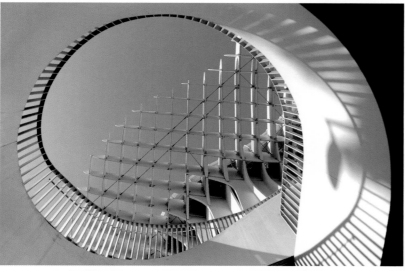

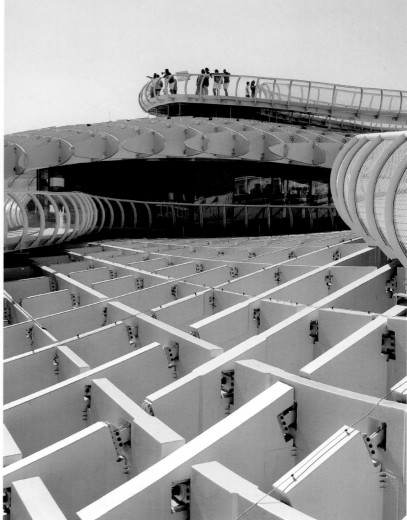

MATERIAL

Steel box truss, Versiweb, Polycarbonate, Timber decking, Recycled PET bottles, Wedelias plant, Straw mats.

Wonder Wall

奇妙墙景观设计

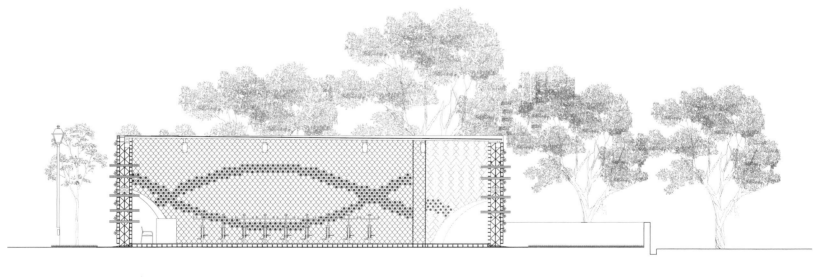

1 SECTION

0 1 2   5   10m

LOCATION

Singapore

AREA

270 m²

PHOTOGRAPHER

Aaron Pocock, C3 Momentum Studio

The Wonder Wall is a zero waste pavilion that reuses materials in a new way to extraordinary functions and delight thus engaging and inspiring the hearts and minds of all to rethink Singapore.

The proposed pavilion seeks to embody the duality between the two realms, with its permeable skin. The undulating web inspires curiosity and amazement as well. At certain angles, the membrane looks almost solid like a wall, and when one moves along the Wonder Wall, a "moire" effect is created due to the double cladding around the structure. When viewed on the perpendicular, the membrane seems totally transparent and merges with the surrounding buildings and landscape.

The zero waste and build ability strategy was developed around two highly rapid deployable and re-useable systems. The first is the main structure, composed of box-truss systems developed for the Formula One Night race and the National Day Parade.

The second is a polymer mesh developed for slope control that has unique attributes that enhance the usability and interaction of the space, the membrane and its landscape system can be re-utilised around Fort Canning for slope and erosion control. The zero waste strategy considered time, materials, cost and the afterlife of the elements. The box-truss system, including the roof, takes a maximum of approximately seven days to deploy. The membrane takes a maximum of approximately three days to install. Overall time frame to complete Wonder Wall erection would be ten to fifteen days. The cellular membrane once taken down can be re-used for the following: Fort Canning Hill's other areas that require slope protection and stabilization; Donate to a nearby country whose village / farmland has been affected by soil erosion from slopes; The steel box-truss once taken down will be re-used in other commercial events along with the future National Day parades.

"奇妙墙"景观是一件纯废品制成的作品，它采用废弃材料，以一种新的方式展现了非凡的功能，因此点燃了人们全部的热情来加入到"反思新加坡"的队伍中。

展馆设计旨在以它具有渗透性的外观，体现两个领域之间的二元性。起伏的网状结构也激发好奇心和惊奇感。从一定角度，隔膜看起来几乎坚硬得像一堵墙。当一个人沿着奇妙的墙移动，隔膜会因为建筑双薄层的结构设计而呈现出"波浪起伏"的样子。在垂直观看时，隔膜看起来完全透明，将周围的建筑和风景融合在一起。

纯废品和构建能力的策略是在两个具有高度快速开展性和可再用性的系统上构建的。第一个系统是建筑的主体结构，由箱形构架系统组成，它是为一级方程式夜间比赛和国庆阅兵建造的。第二个是由聚合物制成，用于控制斜坡，具有独特的属性，提高可用性和互动的空间，隔膜及其景观系统可以在福康宁前坡和侵蚀防护周围被重新利用。纯废品计划考虑到时间、材料、成本和材料的循环使用。包括一个屋顶的箱形构架的系统展开最多需要大约7天的时间，隔膜的安装最多只需要3天左右的时间。完成奇妙墙框架的构建需要10~15天的时间。多孔的隔膜一旦取下可以被再次使用，例如可以用来稳定和保护福康宁其他地区的山坡；捐到附近有饱受坡面土壤侵蚀影响村庄／耕地的国家。钢箱桁架一旦取下，随着今后的国庆节游行将重新用于其他商业活动。

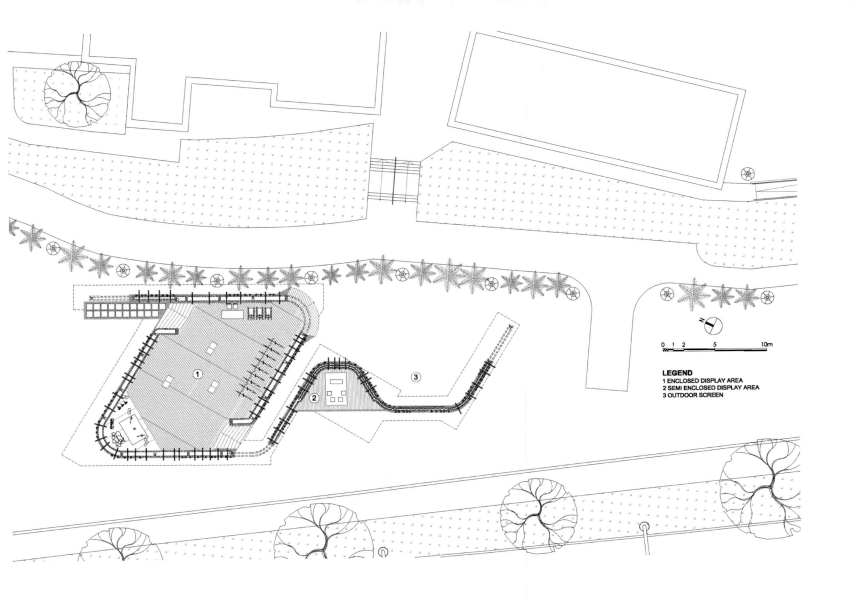

DESIGNER	DESIGN COMPANY	COMMISSIONING AND MANAGEMENT	FUNDING AGENCY	LANDSCAPE ARCHITECTS
Dan Corson	Corson Studios LLC	Regional Arts and Culture Council	Portland Parks	OLIN Landscape Architecture

Mercurial Sky

多变的天空

In downtown urban park devoid of plant material or "nature", splintered ultra low-resolution videos capturing the movement of natural sequences (sparkling water, a breeze moving branches, fire, clouds, pulsating jellyfish) are integrated into the giant glass canopy of Director Park. The low-resolution video was pixel-mapped onto a series of linear LED fixtures that were hidden within the canopy purlin design.

When experiencing the artistic environment directly under the canopy, the low-resolution moving lights provide dynamic and interesting optical experiences while animating the space. From under the canopy, there is no way to comprehend what video imagery is being seen due to the nature and scale of the piece. However, at the perimeter of the park or in the reflections of the adjacent buildings, it is possible to discern the video images when the LEDs are compressed through foreshortening.

So, in the daytime, the clouds and people provide animation for the plaza. After dark, the park becomes alive when the lighting transforms the entire park into a pulsing and dynamic environment while infusing colour, light and movement into the urban plaza.

CANOPY ARCHITECTURE	LIGHTING CONSULTANT	PHOTOGRAPHER
ZGF Architects LLP	Benya Burnett Consultancy	Dan Corson

在缺少植物和自然气息的商业区城市公园中，一块块超低分辨率录像机捕捉着变化着的自然风景（粼粼的水光、摇曳的树枝、火光、云彩，还有游动的水母），并与迪雷克托公园中巨大的玻璃华盖融合在一起。华盖的椽子里安装着一连串线形LED 装置，低分辨率录像机的像素点就映射在其上。

在华盖之下直接欣赏风雅的环境时，低分辨率移动灯光营造出动感有趣的视觉体验，赋予空间生动的气息。在华盖下由于自然环境和玻璃盖大小的原因，没人能预见可以看见什么样的视频影像。但是，在公园周围，或者在相邻建筑的反射下，当LED 灯通过投影缩压时，就可以清楚地看到视频影像。

因此，在白天，广场因云彩和游人变得生机勃勃。天黑之后，当灯光将整个公园的氛围变得跳动活跃时，公园逐渐活力四射起来，与此同时，还为城市广场注入了一丝色彩、光亮与动感。

DESIGNER
Ian McChesney

PHOTOGRAPHER
Peter Cook

Wind Shelters

风形遮蔽景观

Following an open RIBA competition the firm were commissioned to design two rotating wind shelters for Blackpool's newly regenerated South Shore Promenade. The shelters are designed to rotate according to the prevailing wind direction to shield the occupants from the elements. The shape was born out of a distillation of the key required elements: a vane, which will turn the structure, and a baffle that will shelter the inhabitant from the wind. Extensive testing and development work was carried out to establish the performance of the shelters. This culminated in the manufacture of a full sized working prototype. The final shelters are 8 metres tall and manufactured from resilient "Duplex" stainless steel. They sit on 4-metre-diameter turntables, which incorporate a dampers to control the speed of rotation.

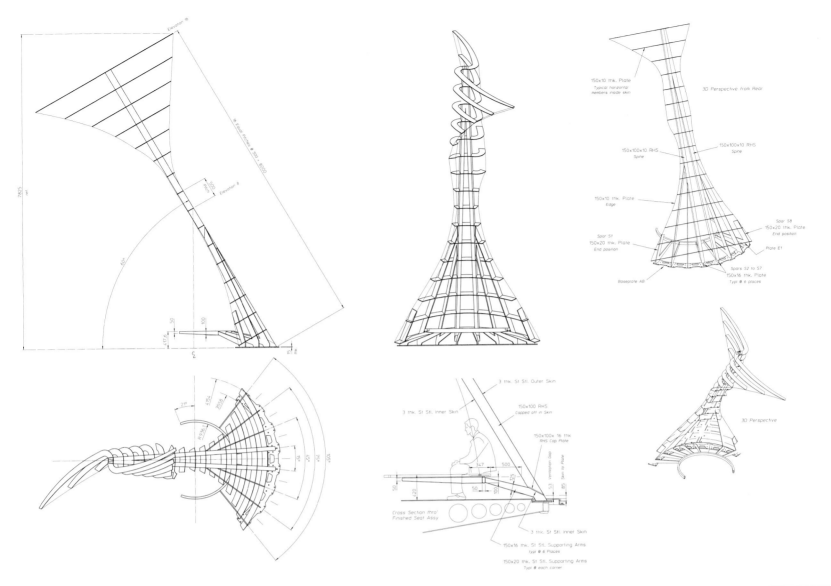

150x10 thk. Plate
Typical horizontal members inside skin

3D Perspective from Rear

150x100x10 RHS
Spine

150x100x10 RHS
Spine

150x10 thk. Plate
Edge

Spar S8
150x20 thk. Plate
End position

Spar S1
150x20 thk. Plate
End position

Plate E1

Baseplate AB

Spars S2 to S7
150x16 thk. Plate
Typl @ 6 places

3 thk. St Stl. Outer Skin

3 thk. St Stl. Inner Skin

150x100 RHS
Capped off in Skin

150x100x 16 thk.
RHS Cap Plate

Ventilation Gap
Skin to Plate

3 thk. St Stl. Inner Skin

150x16 thk. St Stl. Supporting Arms
Typl @ 6 Places

150x20 thk. St Stl. Supporting Arms
Typl @ each corner

Cross Section thro'
Finished Seat Assy

3D Perspective

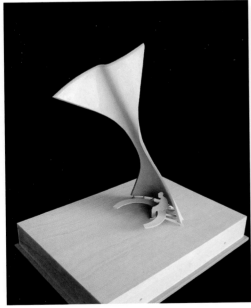

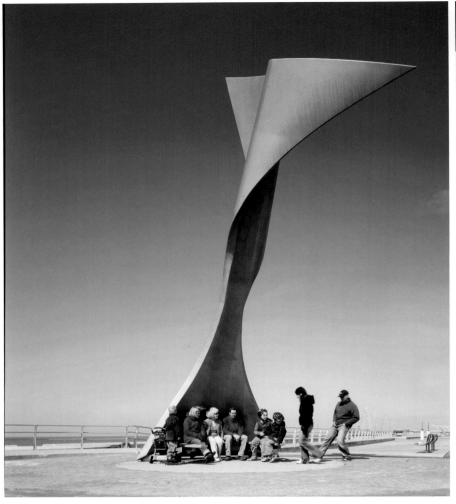

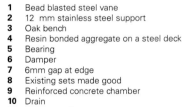

1 Bead blasted steel vane
2 12 mm stainless steel support
3 Oak bench
4 Resin bonded aggregate on a steel deck
5 Bearing
6 Damper
7 6mm gap at edge
8 Existing sets made good
9 Reinforced concrete chamber
10 Drain
11 Concrete piles

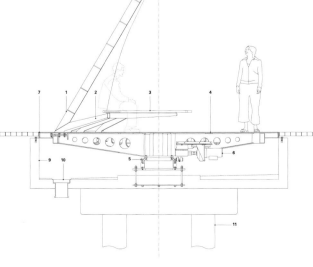

在一个开放的 RIBA 竞赛中，设计公司被委托为布莱克普尔新建造的海滨长廊设计两个回旋的风形遮蔽处。这些掩体设计成旋转状是根据盛行风的方向原理来保护使用者。造型的设计由以下必需要素形成：一个风向标来形成框架；一个防风隔板来遮蔽居民，躲避大风。以广泛的测试和开发工作进行来确保避难所的性能。这是一个全尺寸的工作原型制造。最后完工的庇护所是一座高8米、由弹性的双相不锈钢管制成的建筑。它们坐在4米直径的转盘上，其中包括一个控制旋转速度的减速装置。

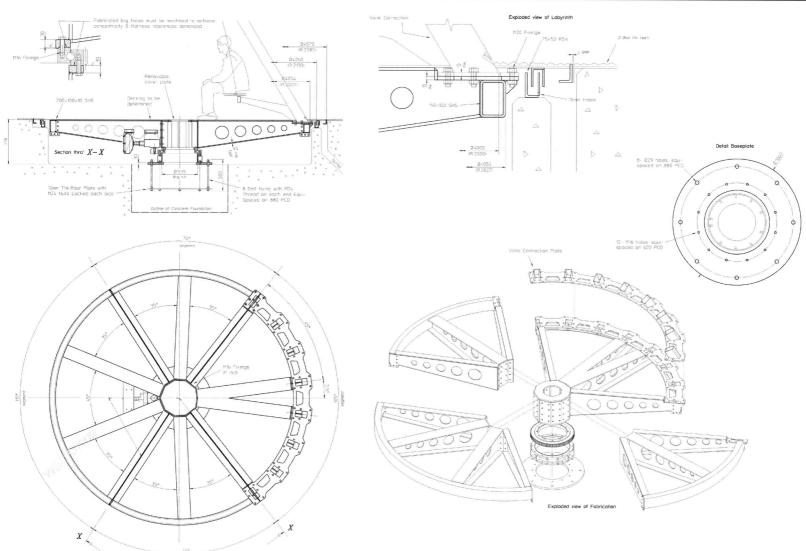

LANDSCAPE ARCHITECT	DESIGN COMPANY	LOCATION	AREA	PHOTOGRAPHER
Michal Riabič, Martin Lepej	Uniform Architects	Slovakia	1,000 m²	B. Boďa, M. Fabian

Molo

莫洛亭

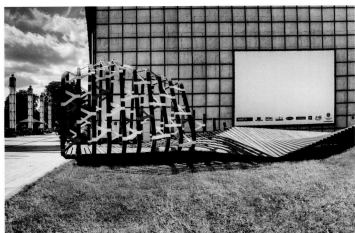

Architecture within the urban space has the power to attract, to guide or even to change the mood of a person. The goal of Molo – culture pavilion for Martin is all of these, but most of all to create opportunities, for a community to meet, for people to relax or for a cultural event to be organised. Uniform Architects is opening to public a long neglected site in the town centre, building an architectural sculpture and creating a space dedicated to culture all at the same time. The open design of Molo pavilion gives everybody the freedom to decide on its use, for a small concert, as an alternative theatre, for a summer reading or movie screening. By merging a strong, dynamic form with a lighting installation Uniform Architects hope to stimulate or give birth to a range of cultural activities and become the culture node of Martin.

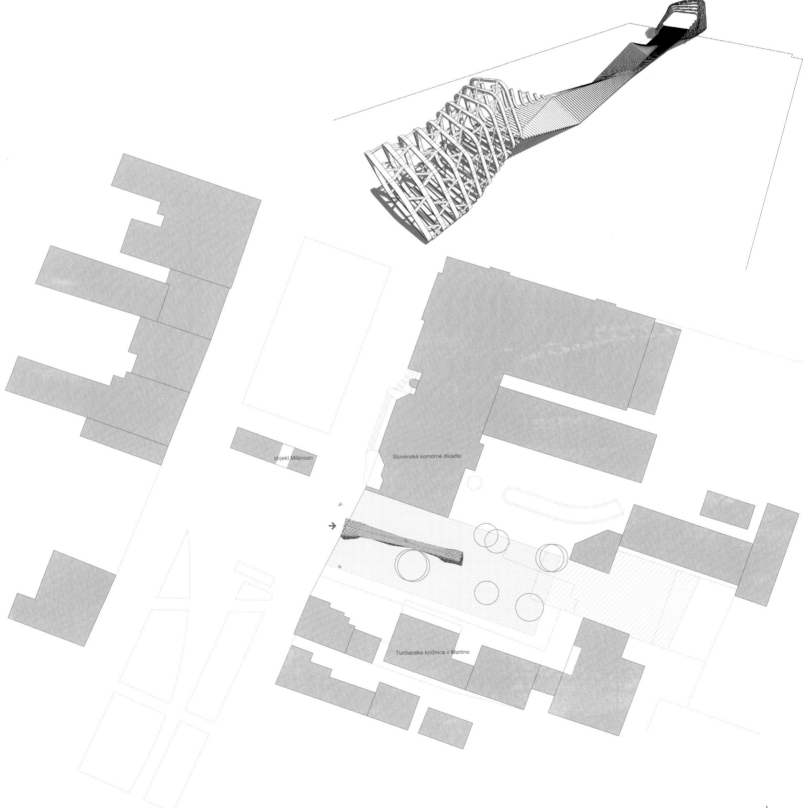

objekt Milénium

Slovenské komorné divadlo

Turčianska knižnica v Martine

　　都市空间内的建筑似乎有着一股魔力，它们引人注目，指引方向，甚至改变人的心情。莫洛亭位于马丁市，是一座文化亭，其目标就是要实现以上所有功能，但最重要的是为人们聚会、休憩，或组织文化活动创造机会。Uniform 建筑师事务所将在镇中心一个向来被人忽视的地方建造一尊建筑装饰雕塑，同时打造一个致力于文化传播的空间，对公众开放。莫洛亭采用开放式设计，人们可以随意改变其用途，如举办小型音乐会，将其用作另类剧院，或者夏季的时候，在这里读书或放映电影。我们将莫洛亭坚固、充满活力的外观与照明设施融合在一起，希望促进该区域一系列文化活动繁荣发展，从而使其成为马丁市的文化节点。

DESIGNER
Andres Silanes Calonge, Fernando Valderrama Garre, Carlos Bañon Blazquez

DESIGN COMPANY
SUBARQUITECTURA

LOCATION
Sergio Cardell Plaza, Alicante Spain

Tram Stop

电车站

Alicante is a city with 400,000 inhabitants on the southeast mediterranean Spanish coast. Over the last years, a new tram infrastructure has been built, using the old rails of the local train. This line connects all the towns of the coast, ending in Denia, a northern city of the province, where ships departure to Ibiza.

This stop is the central stage of a new line of the tram that links the centre of the city to the residential areas of San Juan beach.

The construction of the Tram Stop was an opportunity to bring back a stolen space to the city: to turn a traffic circle into a public space.

Through a fractal access system deformed in each side to avoid the existing trees, the travellers can arrive in a frontal way to the platform in 32 different possibilities.

Over the platforms, two empty boxes (36m long, 3m wide, 2.5m high) create a floating void slightly over the travellers' heads. It matches the size of the train, creating an intermediate scale between buildings and urban elements. There is no difference between structure and envelope, neither between roof and walls. It is an isotropic material in both conception and construction.

The holes reduce the weight as increase the resistance to normal tensions, and equally decrease wind pressures among the surfaces.Light and air pass through, smoothing the shadow and generating a soft breeze in summer months.

At night the boxes are transformed into two giant lamps.

Benchs are spread over the garden close to the vegetation and the paths, creating a public place overlaying the quiet of the seated people and the movement of the people walking.

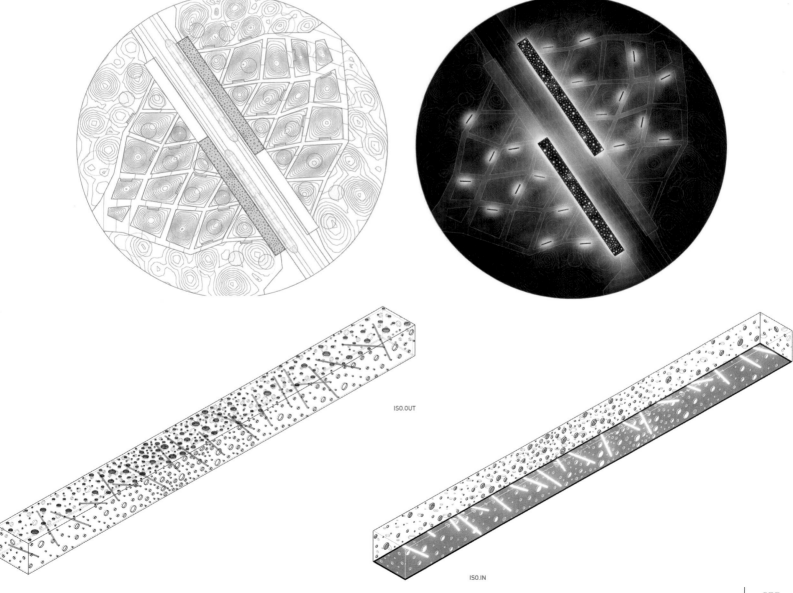

ISO.OUT

ISO.IN

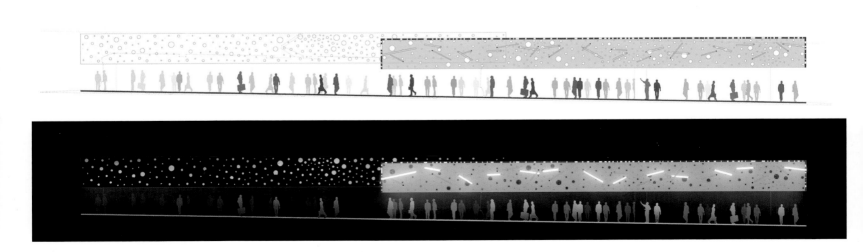

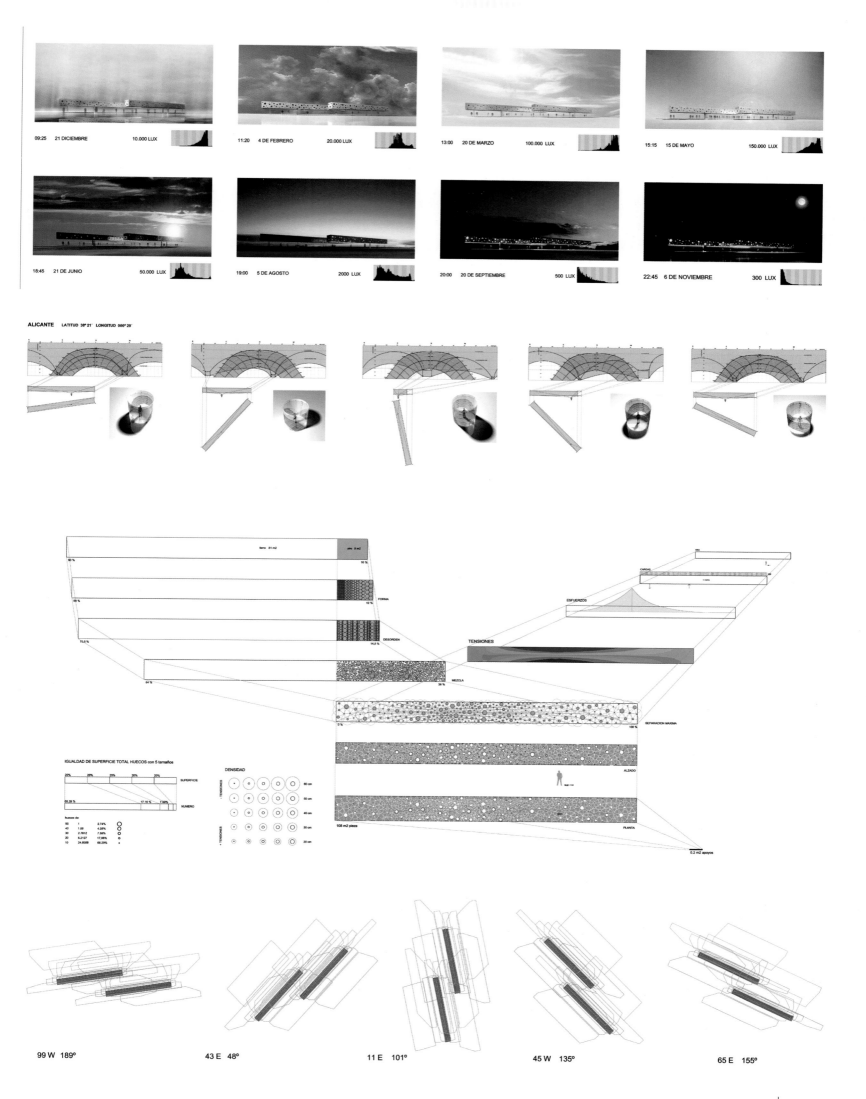

09:25   21 DICIEMBRE   10.000 LUX

11:20   4 DE FEBRERO   20.000 LUX

13:00   20 DE MARZO   100.000 LUX

15:15   15 DE MAYO   150.000 LUX

18:45   21 DE JUNIO   50.000 LUX

19:00   5 DE AGOSTO   2000 LUX

20:00   20 DE SEPTIEMBRE   500 LUX

22:45   6 DE NOVIEMBRE   300 LUX

ALICANTE   LATITUD 38° 21´   LONGITUD 000° 29´

IGUALDAD DE SUPERFICIE TOTAL HUECOS con 5 tamaños

DENSIDAD

99 W 189°   43 E 48°   11 E 101°   45 W 135°   65 E 155°

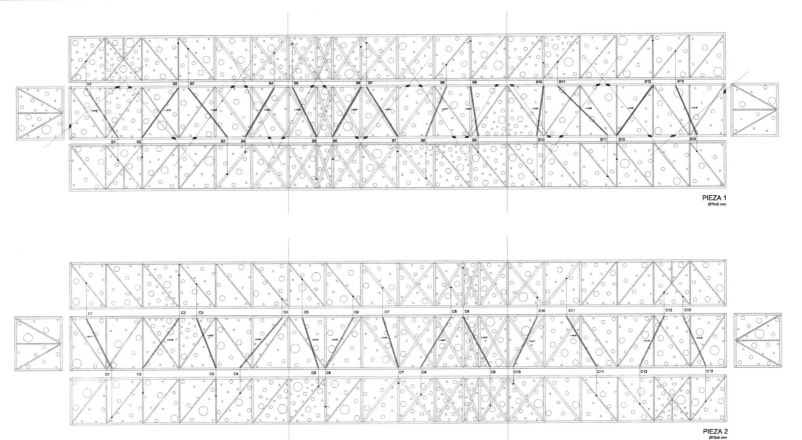

PIEZA 1
Ø70x5 mm

PIEZA 2
Ø70x5 mm

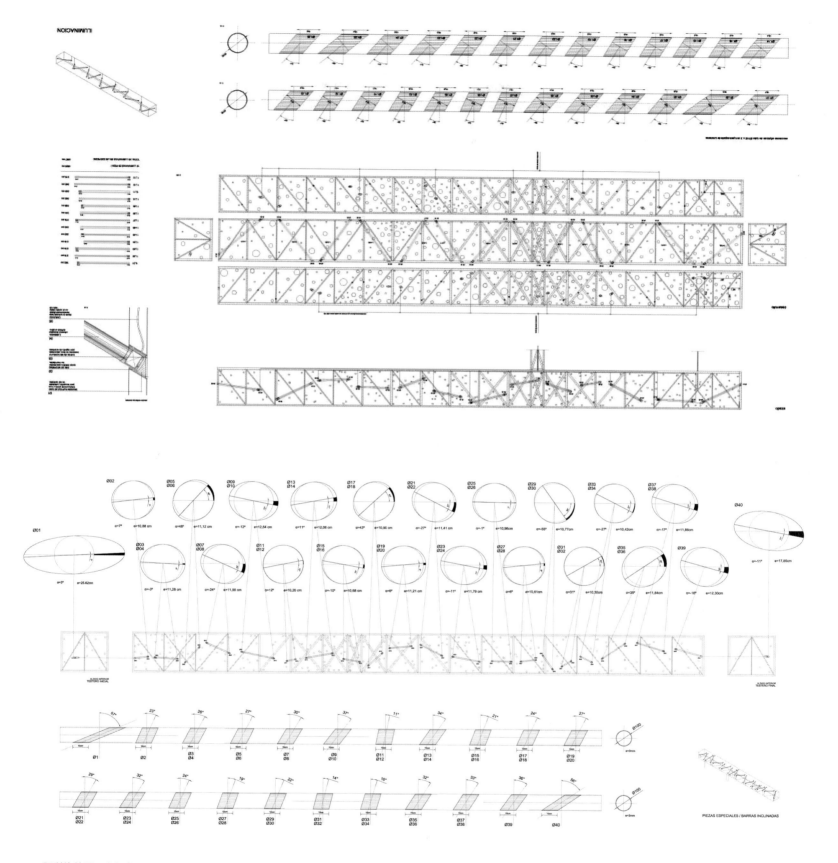

阿利坎特是一座拥有 400,000 人口的城市，位于地中海西南方向的西班牙海岸。在过去的几年中，这里利用废旧铁轨，建造了新的电车基础设施。这条电车线将沿海岸的所有小镇连接了起来，以该省的北方小城德尼亚为终点，从那里可以乘船前往伊比萨岛。

该车站位于新电车线路的中心地段，连接了市中心和圣胡安海滩居民区。

电车站的建设对恢复缺失的城市空间来说是个好机会：环形交通道路圈出公共空间。

电车站的入口系统呈不规则形状，因为要躲避生长在那里的树木，所以入口两边的形状发生了变化，乘客经过该入口系统可以选择 32 种不同的线路到达前方的月台。

月台上方，两个空车厢（长 36 米，宽 3 米，高 2.5 米）在比乘客的头顶略高处创造出悬浮的空间。

它与电车的尺寸很匹配，在建筑物和城市元素之间显得规模适中。

外壳与结构没有什么不同，顶棚和墙壁也没有不同。

电车站的设想和建造用的都是各向同性材料。

车厢上有许多小洞，减少了车厢的重量，增加了法向拉力的阻力，并且同样减少了表面的风压。

灯光和空气穿过，使阴影柔和，并在夏季带来了柔和的微风。

夜里，车厢变成了两个巨大的灯笼。

长椅在靠近植被和小路的花园中随处可见，营造出一个公共场所，掩映着坐着休憩的人们和行走的路人。

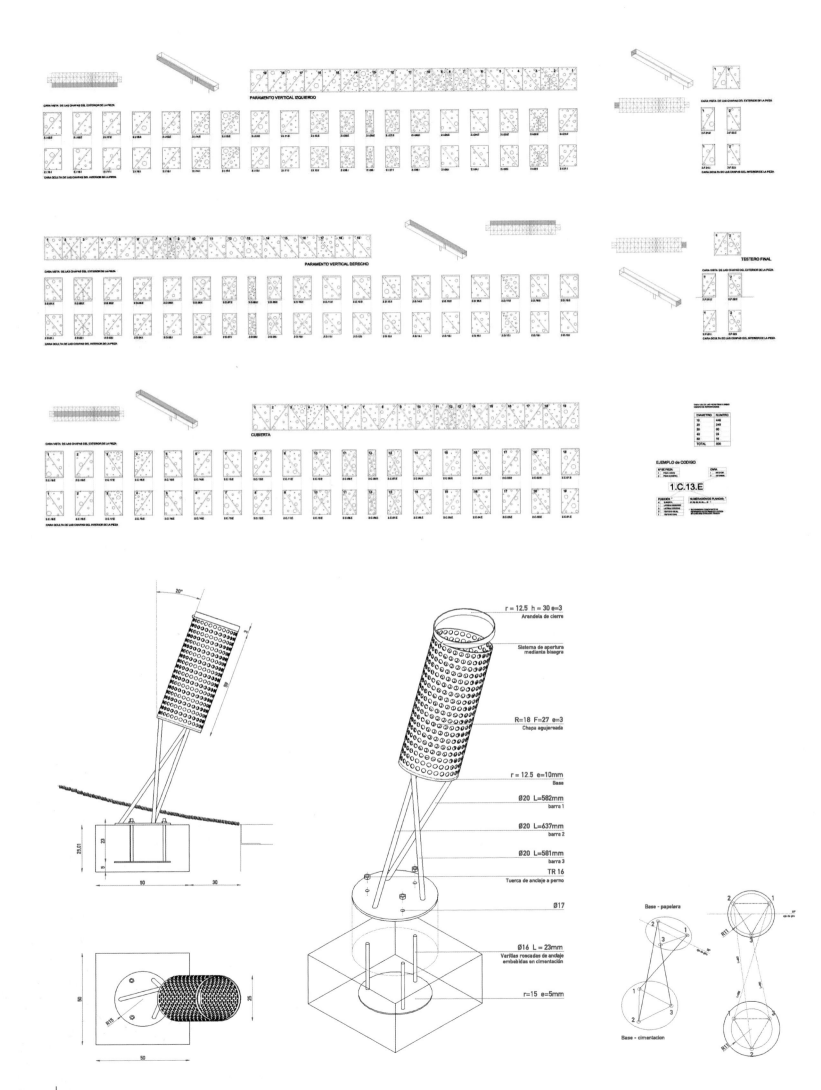

DESIGNER
Miguel A. Alonso del Val, Rufino J. Hernández with Marcos Escartín

DESIGN COMPANY
ah asociados

ENGINEER
VS Ingeniería y Urbanismo, Urgein

LOCATION
Navarr, Spain

Pergolas at Buztintxuri Square

Buztintxuri 广场棚架

Buztintxuri is a new residential area of Pamplona whose urbanisation responds to the basic objective of defining the surfaces of support (infrastructures, pavements, etc.) of an area. In the centre of this new space, pergolas over the pavement create shades and contribute to the integration of built and unbuilt space in a single project.

Several objectives are pursued in this project. First of all, pergolas are designed to contribute to the dimensioning of the wide open spaces in more manageable size. Secondly, these new urban objects establish an intermediate scale between the open spaces and equipment buildings located along its edge, playing the same role as western edge commercial porches. Finally, pergolas generate wide subspaces to shelter from the sun and profit from public spaces.

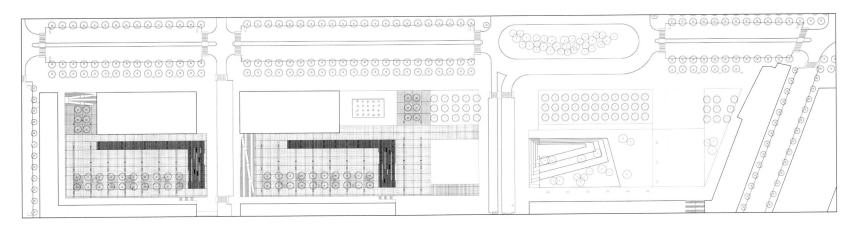

Buztintxuri 是潘普洛纳地区一处新的住宅区，它的城市化建设响应了改造地区大致面貌（基础设施、人行道等）的基本目标。改造后区域的中心，藤架布满人行道，投下斑驳阴影，将虚实空间统一在一个设计里。

改造工程致力于以下几点：第一，棚架的设计要可以开阔空间，又使其处于便于管理的规模；第二，新的城市设施处于中等规模；沿着设备大楼建于大楼与开放区域之间，就像西方的边缘商业门廊那样扮演划分区域的角色；最后，棚架可以提供大片区域让人们远离刺眼的阳光，并提供公共空间的诸多好处。

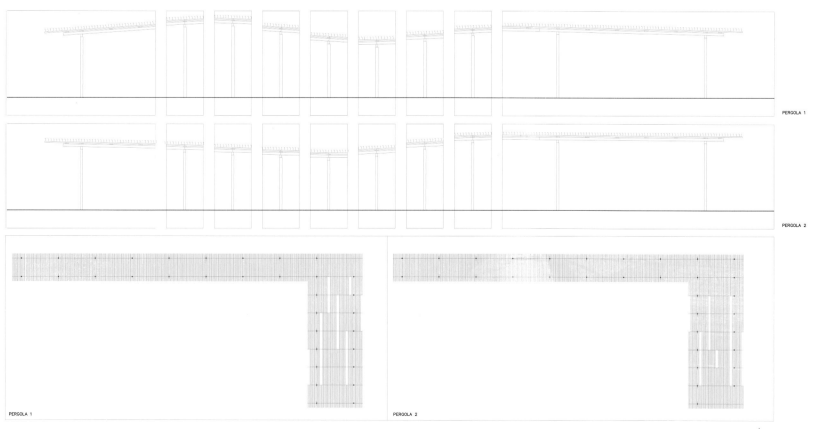

PERGOLA 1

PERGOLA 2

PERGOLA 1

PERGOLA 2

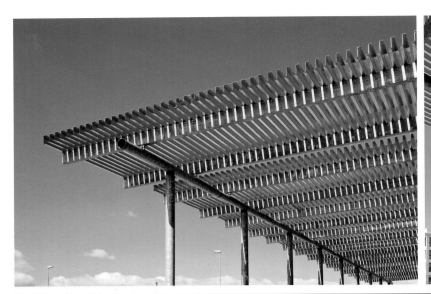

JCDecaux and Mathieu Lehanneur won the call for projects by the Mairie de Paris devoted to intelligent furniture with Escale Numérique, a connected haven of peace available to everyone. Set up on the Rond Point des Champs-Elysées, Escale Numérique is a revival of the underground fibre optic network which is now supplying the capital. "Like the Wallace fountains, which since the end of the 19th century have offered Parisian the free drinking water which was circulating beneath their feet, Escale Numérique allows everyone to benefit, like a real public service, from a high-speed WIFI connection by raising it from beneath the ground."

DESIGNER	LOCATION
Mathieu Lehanneur, JCDecaux	Paris, France

Escale Numérique (Digital Break)

数字休息亭

德高集团和设计师 Mathieu Lehanneur 赢取了由巴黎市政府提倡的智能设施——数字休息亭的开发项目，这是一个人们可以交往的平和的休息场所，向公众开放。数字休息亭位于圆形广场香榭丽舍大道，是由地下光纤网络供电的，就像华莱士喷泉一样，自19世纪末以来，它为巴黎市民供应来自他们脚底下的免费饮用水。如同真正的公共服务一样，高速 WIFI 连接器从地下传到地面上来，数字休息亭向所有公众开放。

DESIGN COMPANY
Oglo

PROJECT TEAM
Emmanuel de France, Arnaud Dambrine, Pauline Rabjeau

CLIENT
Concello de Allariz

La Muneca

人形景观

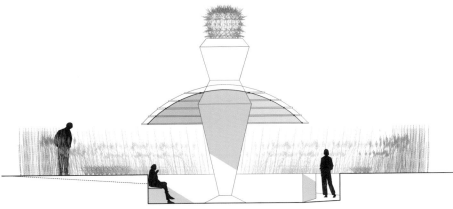

Whatever culture fashion is part of, whatever time and place it refers to, its essence finds its roots in the characteristics of the human body. Only the being's dimensions and proportions matter. Its peculiarities, its flaws, its disadvantages are sublimed by creation and become the assets, the basis, the nourishing roots of fashion. They structure its path, define its substance, and guide its function. Static and still when unworn, fashion transforms itself; it evolves in space once inhabited. It also shelters life. It takes over, welcomes and protects it. Once wrapped around the being, fashion, lifeless until then, comes alive.

PRODUCTION TEAM

Concello de Allariz, Emmanuel de France, Arnaud Dambrine, Romain de Braquilanges, Augustin Caradec, François Leite

LOCATION
Allariz, Galicia, Spain

AREA
250 m²

PHOTOGRAPHER
Oglo (Emmanuel de France & Arnaud Dambrine)

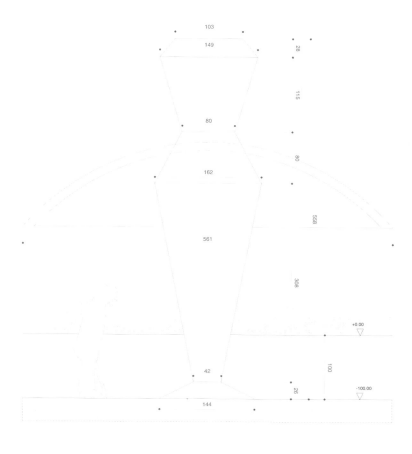

After going across a thin plant screen, people slowly walk down through the flowers and the soil. An internal world is revealed; its soft curves cocoon and guide them. The plants offer as many smells as they do colours. They wrap themselves into space towards the image of their condition, towards the structuring signal. Further below, sheltered from any hazard, they are taken in the centre of the symbol and, unscathed, they contemplate.

Materials :

Floor: Brick, scratched coat, white graval

Muñeca: Steel, sun umbrella canvas, paint

Plants: Panicum Virgatum, Miscanthus Giganteus, Lupinus Polyphylus, Santolina C., Phornium spp, Lotus Berthelotii, Muhlenbergia Capillaris, Penisetum Setaceum, Tagetes Erect, Dimorfoteca, Senecio Cineraria, Chalota Echalote, Buxus Sempervirens, Argyranthemun Blanco, Stipa Tenuissima, Santolina Chamaecyparissas

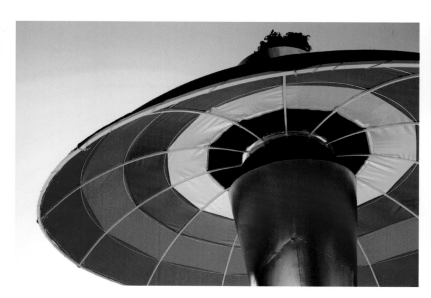

Sketch 1

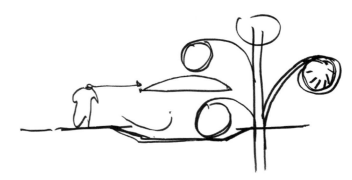

Sketch 2

Sketch 3

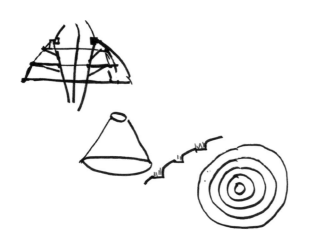

Sketch 4

Sketch 5

Sketch 6

　　任何时尚文化的一部分，是指任何时间和地点，在人体的特征上可以从本质上找到时尚文化的根源。只是尺寸和比例与人体不同。它的特性、瑕疵、劣势是升华的创造，并成为了时尚的资产、基础和营养的根基。他们建造了它的小道，确定它的材质，并导向它的功能。没着装时是安详平静的，而穿上时尚的衣服后，其自身得到了改变；一旦安置好后，它就使空间得到了升华。它还是生灵的庇护所。它接待、欢迎并保护生灵。一旦它被人们和时尚环绕，它顿时就会由毫无生气变得生机盎然。

　　走过一层薄薄的植物屏后，人们慢慢地穿过鲜花和土壤的景观。一个内部的世界展现在眼前，其柔和曲线紧紧包裹着人们，并给他们指引道路。周边植物散发出跟它们颜色一样多的气味。它们根据自身环境的形象和根据结构的标志，将自身笼罩在一个空间里。往更深一步想，它们深思熟虑，远离任何危险，并毫发无损地占据了象征标志的中心。

材料：

地面：砖块、防刮涂层、白沙

娃娃：钢铁、帆布太阳伞、油漆

植物：柳枝稷、芒竹、多叶羽扇豆、绵杉菊、新西兰麻、Lotus Berthelotii、乱子草、Penisetum Setaceum、万寿菊、Dimorfoteca、银叶菊、Chalota Echalote、锦熟黄杨、Argyranthemun Blanco、细茎针茅、Santolina Chamaecyparissas。

A　B　C　D　E　F　G　H　I

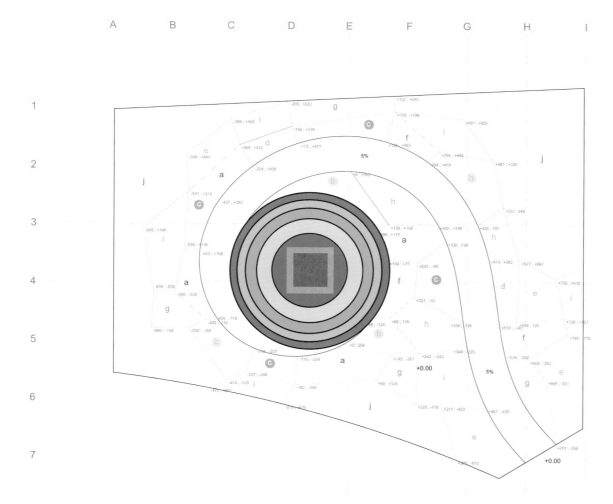

a　MYOSOTIS VICTORIA

b　HELIANTHUS ANGUSTIFOLIUS

c　ARGYRANTHEMUM CALLICHRYSUM

d　ARTEMISIA ABSINTIUM

e　PENNISETUM SETACEUM

f　SENECIO CINERARIA

g　MUHLENBERGIA CAPILLARIS

h　STIPA TENUISSIMA

i　PANICUM VIRGATUM

j　MISCANTHUS GIGANTUS

DESIGNER
David Karásek, Radek Hegmon

DESIGN COMPANY
mmcité

Cortex

树皮庇护所

An originally shaped shelter, the roof of which is extended to the rear wall and made of a trapezoidal steel sheet. The shell that arises is part of the supporting structure of this simple yet stylish shelter. The side-walls are available in self-supporting strip-wood boards for outdoor variations. The construction is also available in steel with glass.

原始形状的庇护所，由梯形钢板制成的屋顶延伸到后墙。升高的外壳是支撑这个简单而时髦的庇护所的部分构造。侧墙透过自支撑的带状木板可以看到外面的变化。镶嵌玻璃的钢材也能搭建这样的构造。

DESIGN COMPANY	CLIENT	LOCATION	PHOTOGRAPHER
Allegory Studio	Festival Arbres en Lumières	Place de Neuve, Geneva, Switzerland	Annik Wetter

Sheltree

广场遮蔽建筑

Sheltree for Arbres en Lumieres (Urban Light and Trees Festival)

Facing an important public square in Geneva, Sheltree is one of the winning projects of the city's annual Light and Trees Festival, which took place in December 2012, during the pre-Christmas period.

The project was designed based on a simple observation. In the summer trees offer shelters to protect from the rain or the sun. In the winter unfortunately, they can't fulfill their role to maintain their ability to shelter as they lose their leaves. In Industrial Design, umbrellas were created to offer shelters to protect from the same things. Sheltree is thus a hybrid project resulting in an installation halfway between a tree and an umbrella, which allows the tree to still fulfill its role of shelter during the rainy winter days.

The umbrella is lit from below the structure by LED light sources. The colours change gradiently, which were chosen according to the natural colour evolution of tree leaves. The structure and the bench are set underneath it. The use of metal allows for the structure not to contrast too much and to rent cars.

广场遮蔽建筑是为城市路灯与树木节设计的。

针对日内瓦一处重要的公共广场，广场遮蔽建筑是城市每年举办的城市路灯与树木节中的一个获奖作品，节日举办在 2012 年 12 月的圣诞节前夕。

这项作品的设计是根据一个简单的观察而来的。在夏天，树木给人们提供遮蔽处来抵挡阳光和雨水。在冬天，不幸的是它们再也不能继续扮演遮蔽处的角色，因为树叶都已凋落。在工艺设计中，伞也是被造来保护人们免受暴晒和雨淋的。广场遮蔽建筑因此是一个混合项目，成为将树木和雨伞结合的设施，这使得树木可以在下雨的冬天继续扮演其遮蔽处的角色。

伞状结构的下方被 LED 光源照亮。颜色呈梯度变化，颜色的选择是根据树叶自然颜色的变化。支架座椅设置在它底下。使用金属，建筑不会反差太大，方便车辆停靠。

DESIGN COMPANY	MATERIAL	LOCATION	PHOTOGRAPHER
FTL DESIGN ENGINEERING STUDIO	PTFE glass fabric, galvanized cables, stainless steel	Scottsdale, AZ, USA	FTL Design Engineering Studio

Skysong

天空之歌

FTL developed an iconic signature element called "SkySong" which is a 50,000-square-foot campus marker and shade structure straddling two intersecting roads and four plazas. The shaded plazas will become the heart of the new centre with retail and restaurants located under the passively cooled structure. The PTFE glass fabric structure opened in April 2010.

he tensile structure uses a series of cable suspended "tensegrity" steel trusses which create a shade cover for the campus courtyard. A-frame masts support these steel trusses with an array of cables using rotational symmetry, suggesting movement of form and light across the plaza. FTL developed the lighting with the lighting designer Mathew Tanteri using a combination of programmable LED fixtures to illuminate the A- frame masts and metal halide lighting for the fabric surfaces.

The structure has a surface area of approximately 4,000sqm, consisting of a PTFE glass fabric (Sheerfill I) with both galvanised cables at the mast tops and stainless steel fittings at pedestrian levels. The basic structure is a series of four A-frame masts which support both the fabric with rings pulling up and down. The masts support a series of four trusses which are held in position with steel cables, providing coverage to the corners of the plaza. Between the trusses and the fabric rings the membrane is tensioned. The structure utilises rotational symmetry, which allows the surface to appear as a free form shape, but is created from four similar modules. The A-frames were installed first, followed by the four trusses which where suspended between. With the trusses stabilised the fabric fields were installed and tensioned. Raising the trusses into their final position applied prestress to the structure and completed the installation.

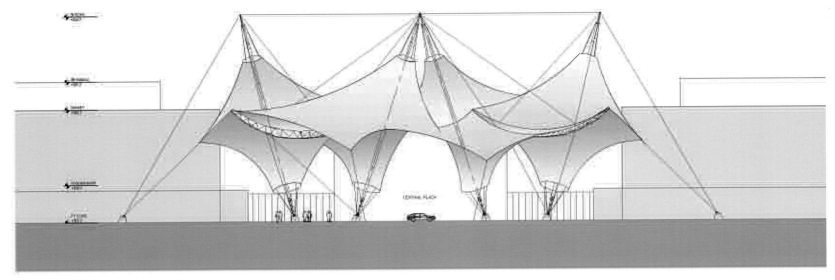

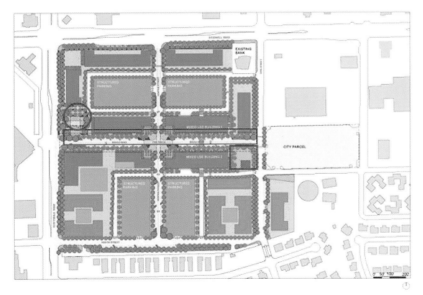

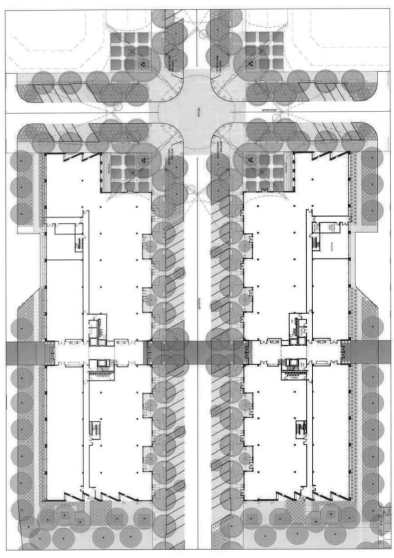

FTL 设计公司设计了标志性的识别元素"天空之歌"，这是一个 4645 平方米的校园标记和跨交叉路和四个广场的遮阳结构。林荫广场将会成为零售店和餐馆新中心的要地，零售店和餐馆被设置在凉爽的结构下。在 2010 年 4 月，聚四氟乙烯的玻璃纤维织物结构已经完成。

拉伸结构采用了一系列电缆悬浮的张拉的钢桁架，创建一个绿荫覆盖的校园庭院。人字桅杆支持这些钢桁架，钢桁架上的绳索轴对称组合设置，提示其在广场的形式和光的运动。FTL 设计公司推出的光照设施是照明设计师 Mathew Tanteri 编程的 LED 灯具组合，照亮一架桅杆和金属卤化物照明表面的织物。

该结构的表面大致由 PTFE 玻璃纤维织物（sheerfill I）构成，镀锌电缆设置在桅杆顶部不锈钢配件安装在人行道上。基本结构是一个系列的四帧的桅杆，同时支持物的环上下提拉。此外，人字桅杆支持一系列的四个桁架，在钢丝绳的位置，覆盖了广场的角落。在桁架和环之间环膜紧紧拉伸。利用结构的旋转对称性，使表面呈现一个自由形状，由四个相同的模块创建。框架首先安装，其次是四个模块悬浮在框架之间。固定环膜的构架被设置和拉伸。提高构架到最终位置来施加预定力量，并完成了装置。

ARCHITECT
Miloš Milivojević, M.Arch.

STRUCTURAL ENGINEER
Milan Zlatanović

LOCATION
Tašmajdan Park, Belgrade, Serbia

PHOTOGRAPHER
Miloš Milivojević

Strawberry Tree Black

太阳能充电设施

A Serbian architect, Miloš Milivojević, designed a new public solar charger for mobile devices Strawberry Tree Black invented by the Strawberry Energy Company.

The charger is conceived as an artificial tree which transforms solar energy into electricity, thereby joining the surrounding trees in a common struggle for the planet richer in oxygen. The steel tree-like construction is large but thin and artistic, which makes it look like a sculpture. The square surface at the top of the solar charger is covered with nine thin-film glass solar panels and nine supporting glass panels which simultaneously acts as a roof, protecting against bad weather conditions. The Strawberry Tree Black acts as a constant reminder of the insufficiently exploited potential of the Sun's energy.

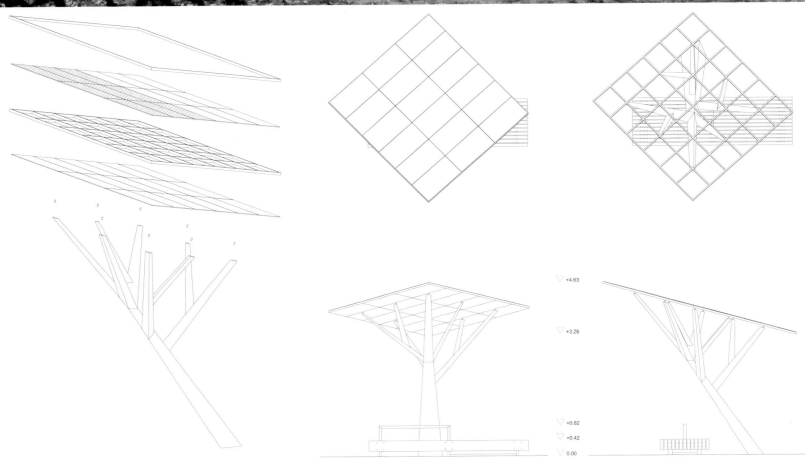

塞尔维亚的建筑师 Miloš Milivojevic 设计了一个为草莓能源公司开发的移动设施——新的太阳能充电设施。

该充电设施是将太阳能转化为电能的人造树，从而与周围的树木一起，努力为地球提供更多的氧气。钢质树形构造，大却轻薄，富有艺术感，这使得它看起来就像一个雕塑。充电设施顶部的方形表面被 9 块太阳能电池玻璃薄板覆盖，同时玻璃也作为防护顶，使设施免遭恶劣天气条件的破坏。太阳能充电设施可以作为一个对太阳能潜能开发不足情况的持续提醒。

DRINKING
FOUNTAIN

饮水台

Periscopio Drinking Fountain
Periscopio 饮用喷泉

G Public Fountain
G 形公共饮水喷泉

DESIGNER
Diana Cabeza

PRODUCTION AND COMMERCIALIZATION
Estudio Cabeza

DEVELOPMENT TEAM
Diana Cabeza, Diego Jarczac, Alejandro Venturotti

Periscopio Drinking Fountain

Periscopio 饮用喷泉

Urban drinking fountain based on the re-use of a drainage pipe. It includes a drainage grate. An interesting complement for the Rehué system.

Materials: Cast iron with polyester powder coating

Fixation: Embedded on in-situ concrete base

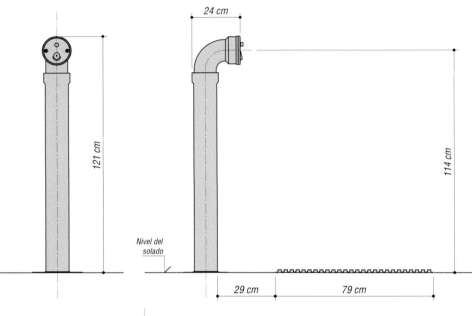

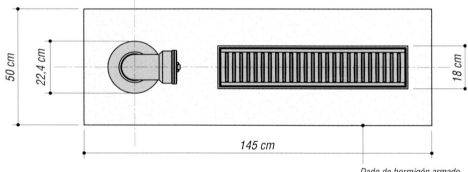

Dado de hormigón armado moldeado un situ, no provist por Estudio Cabeza (ver ficha de Instalación).

城市饮用喷泉是在旧的排水管基础上设计的。
它包括一个喷水口。
是对 Rehué 系统的巧妙补充。

铁质构造刷以聚酯粉末涂料，
在原处嵌入混凝土底层。

DIMENSIONS

W 22cm x D 120cm x H 121cm

DESIGNER
Miloš Milivojević, M.Arch.

LOCATION
Belgrade, Serbia

PHOTOGRAPHER
Miloš Milivojević

G Public Fountain

G 形公共饮水喷泉

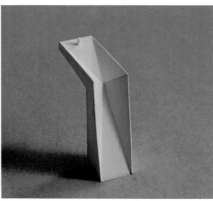
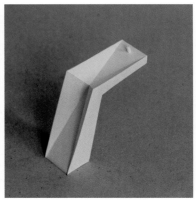

G PUBLIC FOUNTAIN

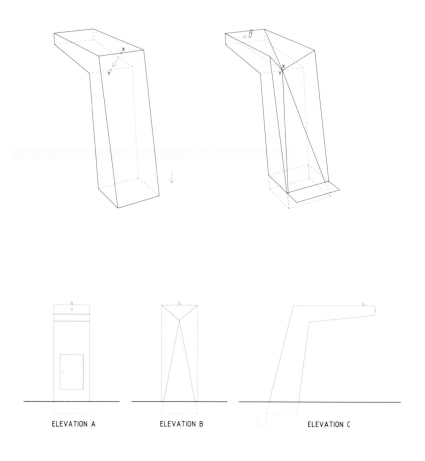

ELEVATION A   ELEVATION B   ELEVATION C

BASEPLAN

SECTION 1-1

1. STAINLESS STEEL CONSTRUCTION
2. TAP
3. INSPECTION OPENING
4. INSPECTION OPENING
5. INSPECTION OPENING
6. STEEL ANCHOR
7. REINFORCED CONCRETE FOUNDATION
8. DRAIN
9. VALVE
10. WATER PIPE

G Public Fountain is the winning proposal in the Competition for Belgrade New Public Drinking Fountain organised by City of Belgrade, Secretariat for Utilities and Housing Services and Belgrade Waterworks and Sewerage. The drinking fountain is designed in the form of a Cyrillic letter G (Г), hence its name. Thanks to its shape, the G Public Fountain is accessible to various users, from adults and children to disabled people and even dogs! The main volume is sunk and dinamysed with two symetrical folded plates which direct the flow of water. The geometry of the folded plates is making a small water game by narrowing and expanding the form of waterfalls. The drinking fountain is completly made from stainless steel.

G 形公共饮水喷泉是"贝尔格莱德新公共饮水喷泉竞赛"上的优胜方案，该竞赛由贝尔格莱德市秘书处的公用事业，住房服务和贝尔格莱德供水系统、排水设备所举办。多亏了这个形状，G 形公共饮水喷泉备受各种各样的使用者的青睐。从成

人、儿童到残疾人士，甚至是宠物狗！它的主体是凹陷的，平均分成两个对称的折面，使水流顺着凹槽流下。以瀑布的形式先缩小后放宽的几何状折面，让水流做了个小游戏。饮水喷泉全部都是用不锈钢制成的。

Mark A. Reigelman II

Mark A. Reigelman II is an internationally recognised Brooklyn-based artist specialising in site-specific product design, installations and public art. His intent is to reshape aspects of the urban landscape in order to provide fresh interpretations of local history and character. By favouring the process of research and exploration over promoting personal artistic agendas, Reigelman has a unique body of work poised between abstraction and literal representation, guided by a clear conceptual foundation. By questioning expected qualities and identities, his work is able to convey ideas, generate conversations, and promote novel, yet enduring, engagements with its audience. Reigelman's work has been exhibited in public spaces, galleries and museums across the country including the Museum of Art and Design (NYC), Museum of Modern Art (NYC), Oklahoma City Museum of Art, Cleveland Museum of Art and the Shanghai Museum of Glass in China. For the past three consecutive years Americans for the Arts have recognised Reigelman's public art installations (Manifest Destiny!, White Cloud and Wood-Pile) as being among the top 50 public art installations in North America. Mark is a member of the American Design Club (AmDC) and co-founder of the New York-based creative collective Art Stars and bi-coastal public art collaborative Chapman/Reigelman.

Tonkin Liu

Tonkin Liu is an award-winning architectural practice, whose work encompasses architecture, art and landscape. They offer forward-thinking clients a design that is finely tuned to the place it is sited, the people who will occupy it, and the culture that surrounds it at the time.

This emphatic search for new beginnings is set out in their book "Asking, Looking, Playing, Making", published in 1999. The unique storytelling methodology searches for archetypes that will inform the process of design from inception to completion, giving the project a lasting resonance.

Each project embodies the relationship to nature. Some projects celebrate changing weather and seasons, some evoke the power of nature as symbols, whilst others emulate form and performance, using lessons in nature to inspire pioneering construction techniques. Their preoccupation with nature informs the design process, whether through biomimicry or by using the elements nature generously gives human beings for free.

They are interested in doing what they have not done before and their aim is always the same, to satisfy the mind and touch the heart.

Manuel Ruisánchez

Manuel Ruisánchez set up Ruisánchez Arquitectes, in its present form, in 1997. The projects of Ruisánchez Arquitectes are of a vast variety of fields and many different scales.

Since it was set up, the studio has won numerous competitions of landscape, city planning and architecture.

In the area of landscape, Ruisánchez Arquitectes has executed a great number of projects of a varying nature and scale. The studio's activity focuses on the transformation of the existing urban fabric: on the design of public space, both in urban and suburban areas; and on the regeneration of degraded areas. Interventions are carried out in natural landscapes, and special solutions are sometimes sought for "border cases" where urban growth requires a different approach.

FÜNDC

FÜNDC, Spanish architecture and urban planning office, FÜNDC = Fusion & Union Needed by  Disciplines of Creation, as they understand that any product, "from a book to a city", needs to be created with a multidisciplinary approach through the coordination from start of all technical and artistic fields necessary.

FÜNDC has been awarded in numerous occasions with architecture and urban planning prizes, and has developed other large-scale urban projects in The Netherlands (Amsterdam and Eindhoven) and France (Lille). Other realised work to be emphasised is the R13 building in China, built inside the Park of Architecture of Jinhua City; project curated by Chinese artist Ai Weiwei and Swiss architects Herzog & de Meuron.

Ignacio Ciocchini

Ignacio Ciocchini is an award-winning Industrial Designer specialised in street furniture products, urban design, and public space design. He is Vice President of Design at Bryant Park Corporation, 34th Street Partnership, and Chelsea Improvement Company, three leading Business Improvement Districts in New York City. As a design consultant, Ciocchini works with government agencies, transportation authorities, real estate developers, and architecture firms that have an interest in adding custom products to their projects. Ciocchini's designs have been included in exhibitions at the Copper Hewitt National Design Museum in New York City, The Autodesk Design Gallery in San Francisco, and The Guangzhou Design Week in China.

mmcité

mmcité is not only a supplier of high-quality street furniture, the company is also a partner to all those who want to create something special within public spaces. Mayors of cities of all sizes in the mountains or coastal areas, architects of small teams and large design institutions, construction companies of local or transnational importance – they address all of them with the aim of achieving the perfect project.

At the beginning, there is always a designer sketch, a mere intention. The strong team of experienced and educated professional create strong, high-quality products. Efficient functionality, careful processing and affordable cost are the main parameters that they monitor throughout the process. Modern design and distinctive expression represent a constant standard of mmcité.

They combine the very best materials, which are further continuously tested. They draw from two sources; on the one hand, from the years of their experience, and on the other, from constant efforts to upgrade the materials. Only the best will pass through a sieve. Function, durability and of course price are what matter most. City public spaces are fascinating places where people meet history.

Moradavaga

Moradavaga is a collective born from the collaboration of architects Manfred Eccli and Pedro Cavaco Leitão. Working since 2006 around the issues of vacant spaces, derelict buildings and the activation of the public realm, the scope of its activity lies in the intersection between architecture, art and design. Through the use of "performative objects" or active interventions such as social workshops, ideas competitions or architectural happenings, Moradavaga's main goal is to bring a little bit of surprise into people's everyday life, making them look at space from different perspectives and appreciate its potential while providing new tools for them to explore and activate it.

STORE MUU design studio

1997 Unite "STORE MUU" for design the surroundings.
2007 Established "Store MUU design studio" in Tokyo.
2008 Registered office of architect.

Ippei Kimoto:
Master of Architecture, Hosei univ., Tokyo. 2000-2006
Nikken sekkei ltd. Registered architect.

Daisuke Ito:
Master of Engineering, Hosei univ., Tokyo. Registered civil engineer.

Keha3

Three men – Ville Jehe, Margus Triibmann and Tarmo Luisk – established OÜ Keha3 in 2009. Their idea was to create in Estonia the first design company managing entire production and sales chain. The areas of activity of Keha3 include design of own products; production and sales, design management and provision of design services. Currently the company employs six people.

The three men established Keha3 to make the world a better place. "We have ambitions, courage and experience to create big things: change your and our own fantasies, wishes and needs into tangible items." Many of their products are designed for outdoor use. If an item is capable to endure such conditions, it is capable to attend them also in much friendlier conditions. All their products are made in belief that they are the right things!

Springtime

Springtime is a creative force that creates exciting, sustainable and paradigm-  shifting products, brands and experiences in the field of sustainable mobility, sports equipment, interior, public design, consumer electronics, juvenile products and brand development. They are based in Amsterdam and have their tentacles all over the globe. Springtime (1995) provides strategic consulting, design research, concept design and industrial design. They offer integrated design – a synergy of communication, graphic design, interface design, and product design. In order to do so, they work together closely with specialists in all fields. This way, they are able to provide their clients with exactly the innovation they need – from strategy to production. Brand, product and media are their ingredients for successful innovation.

Over the years, Springtime has been awarded with many design awards, both in The Netherlands and internationally. Springtime is member of the Dutch Designers Association (BNO).

Jason Flannery

Jason Flannery joined the Forms+Surfaces Design Studio in 2004. He holds a BFA in Industrial Design from the Rhode Island School of Design. Some of his contributions to the Forms+Surfaces' product line include the Bike Garden, Duo Bench, and Knight and Apex families of products.

LODEWIJK BALJON landscape architects

LODEWIJK BALJON landscape architects' work demonstrates the importance of the context, both time and place. The history of the site is a starting point for the future. Enhancing the character of the place, combined

with an analytical treatment of the programme, forms the basis of the design.

The activities of the firm range from the garden to the city. Interest in the complexity of the situation and the programme, and inspiration from the arts and crafts aspect is the binding component in the projects.

Analysing the given situation and programme, researching the context and its potential, and sketching the consequences, results in an integral plan, where all possible aspects fall perfectly together.

BIG – Bjarke Ingels Group

BIG – Bjarke Ingels Group is a leading international partnership of architects, designers, builders and thinkers operating within the fields of architecture, urbanism, research and development. The office is currently involved in a large number of projects throughout Europe, North

America and Asia. BIG's architecture emerges out of a careful analysis of how contemporary life constantly evolves and changes, not least due to the influence of multicultural exchange, global economic flows and communication technologies that together require new ways of architectural and urban organisation. BIG is led by partners – Bjarke Ingels, Andreas Klok Pedersen, Finn Nørkjær, David Zahle, Jakob Lange, Thomas Christoffersen and Managing Partners, Sheela Maini Søgaard and Kai-Uwe Bergmann – with offices in Copenhagen and New York. In all their actions they try to move the focus from the little details to the BIG picture.

Studio MA.A&D

Studio MA.A&D is a design and architecture studio runned by Marcus Abrahamsson from Stockholm, Sweden that works in different constellations and collaborations within the design and architectural field. His work spans from product design, mainly focusing on public space furniture and spaces to larger scale architecture focusing on the development of materiality and production.

He firmly believes in design and architecture coming from the needs of people and never works with fabrications of trends and pleasing current styles. Further more he also believes that at this moment the needs of people equal the needs of the environment. No question is greater than sustainability for our generation.

TORAFU ARCHITECTS

Founded in 2004 by Koichi Suzuno and Shinya Kamuro, TORAFU ARCHITECTS employs a working approach based on architectural thinking. Works by the duo include a diverse range of products, from architectural design to interior design for shops, exhibition space design, product design, spatial installations and film making. Amongst some of their mains works are

"Template in Claska", "Nike 1Love", "House in Kohoku", "airvase" and "Gulliver Table". "Light Loom (Canon Milano Salone 2011)" was awarded the Grand Prize of the Elita Design Award. Published in 2011were the "airvase book" and "TORAFU ARCHITECTS 2004-2011 Idea + Process" (by BIJUTSU SHUPPAN-SHA CO., LTD.) and in 2012, a picture book titled "TORAFU Small City Planning" (by Heibonsha Limited).

AND-RÉ

AND-RÉ is multidisciplinary Portugal-based office, with a global international action, dedicated to the study, research and practice of architecture, design and art. The firm is directed by the partner architects Bruno André and Francisco Salgado Ré. The genesis of their creative identity lies in a spirit

of constant demand. Their creations are the result of hard work, inspiration, will and encouragement of active minds that doubt and question the paradigms of reality, imagining alternate contexts, allowing them to bend the corners of innovation and go beyond.

AND-RÉ approaches main issues with fresh spirit, innovative methods and open-minded personality. The office has been strengthening its position in the emergent architecture panorama, adopting an active critic posture before established contexts. Their goal is to pursue and maintain positive differentiated responses, in a time of fast paradigm changes. Currently the firm works on various programmes, of different types and scales. Their mission is to achieve high levels of quality, innovation and creativity in all projects.

"We love architecture. We love design. We love art and their relatives. We do it with pleasure!" – AND-RÉ

Daily tous les jours

Daily tous les jours creates new collective experiences for public spaces. They believe in participation – empowering people to have a place in the stories that are told around them. Coming from the fields of interaction design and narrative environments, they research ways to interact and tell stories. Their projects bring magic to everyday places, behaviours and objects. These experiences take many shapes: urban interventions, exhibitions, products, spatial design, events, software applications and films.

Their work has won numerous international recognitions including prizes from 2013 Interaction Awards and Core 77 Design Festival. They recently won the first prize at the Unesco Shenzhen Design Awards.

NIPpaysage Landscape Architects

Since its inception in 2001 by five Université de Montréal graduates (Mathieu Casavant, France Cormier, Josée Labelle, Michel Langevin and Mélanie Mignault), NIPpaysage is a leader of a new wave of landscape architects.

NIPPAYSAGE

After having participated in several international projects for Martha Schwartz Inc. and Hargreaves Associates, NIP's founders joined forces and settled in Montreal to bring a new and energetic vision to the field of landscape architecture. NIPpaysage currently has a team of twelve collaborators.

Mikyoung Kim Design

Mikyoung Kim Design is an award-winning international landscape architecture group whose work focuses on merging sculptural vision with the urban landscape. As principal and design director, Mikyoung Kim has brought her background in sculpture and music, as well as her design vision as a landscape architect, to the firm's diverse work. Projects are comprised of designs that meld sustainable initiatives with urban form to develop engaging and poetic landscapes. Over the past fifteen years, Mikyoung Kim Design has been involved in urban scaled projects throughout the U.S. and in East Asia and the Middle East.

Since the firm's inception, the work of MYKD has received critical acclaim winning multiple national awards from the American Society of Landscape Architects, the American Institute of Architects, the Urban Waterfront Centre, The Harvard Design School, and the International Federation of Landscape Architects, as well as awards from Boston Society of Architects and the Boston Society of Landscape Architects. Recent work has been featured in numerous publications, including Architectural Record, the New York Times, the Washington Post, Sculpture, Dwell, Surface Magazine Landscape Architecture, Land Forum, Garden Design, Interior Design, Pages Paysages, and a monograph of the firm's work in Inhabiting Circumference.

Alexandre Moronnoz

Born in 1977 in Bourg-en-Bresse, following an applied arts foundation course at ENS Cachan, he went to ENSCI, Les Ateliers (Ecole Nationale Supérieure de Création Industrielle, Les Ateliers, Paris), from where he graduated in 2003. Since then he has worked as a self-employed designer in Paris. He alternates between research projects, commissions and collaborations, enabling different environments to be explored.

His first collaborations with Delo Lindo, Radi Designers, Elium Studio and Cédric Ragot, led him to alternate work with his own commission and research projects, to develop a contextual design where meetings are an opportunity to invent new object or space proposals together. The connection between architecture and design in his personal research and his way of envisaging projects, leads him to tackle the object like architecture, where questions of context and environment are integrated with the design.

Studio Weave

Studio Weave is a young energetic design practice

Studio Weave

working on art and architecture commissions across the country. They are currently working on a diverse set of projects, from an artist's studio on the west coast of Scotland, to a number of high-profile public spaces across London, and they recently collaborated with contemporary dancers on the set design for an immersive performance.

They aim to create places through playing into and exploring the narratives of spaces. It is important to them that their projects grow from the unique aspects of a place, through its physical and geographical qualities, its use both currently and historically, as well as its myths and legends. They particularly appreciate the quirky and eccentric characteristics that make somewhere distinctive. They have used stories as a way to personify landscapes and design proposals, and have even realised designs that fictional characters have designed themselves!

Karres en Brands landscape architecture + urban planning

Sylvia Karres and Bart Brands founded Karres and Brands landscape architects in 1997. Since then the firm has worked on a wide variety of projects, studies and design competitions in The Netherlands and internationally. Their work encompasses all levels of scale and all aspects of the public domain,

from small-scale interventions such as street furniture, detailing of public spaces, gardens and parks, to urban development plans and strategic assignments.

Every project is approached as far as possible from a diversity of scales and disciplines, without assuming any presupposed hierarchy: after all, every level of scale can potentially exert a major influence on the final plan. This also results in various forms of partnership with representatives of other disciplines from outside of the firm, such as architects, artists, social scientists, ecologists or technicians.

Solovyovdesign

Solovyovdesign is a studio founded by industrial designers Maria and Igor Solovyov, based in Minsk, Belarus. They do product design. You may see conceptual and realised projects in their portfolio. They do furniture, lighting, electronics, accessories and other products. They have been working with clients from Belarus, Russia, USA, France and China. At Solovyovdesign, the purpose is to create useful, emotional and beautiful objects with the finest craftsmanship.

Damien Gires

Based in Paris, Damien Gires actually answers for architecture, scenography, interior design, furniture design, industrial design, and lighting design. His culture is to work on the consciousness of the moment, in order to create as many forms as functions.

Spaces, volumes and forms are developed with economy and simplicity, without loosing the expression, the poetry and the fonctionnel sense of the material.

It remains the game and the surprise...

Nichola Trudgen

Nichola Trudgen has recently graduated with a first-class honours Bachelor of Industrial Design in Auckland, New Zealand. She is a very creative and outgoing person. She has a very strong will to succeed and can apply herself to many different situations but during the process it is crucial that she is having fun and an open mind. Design and being creative brings her so much joy. Nichola Trudgen loves to create and design things that benefit others in some way and see them enjoying what she has made.

D.A.R. Design

D.A.R. Design is a specialist design and manufacturing company that produces creative metalwork and public art as one of their commissions. The range of work covered includes small architectural details such as grills and railings through to a 40-metre pedestrian footbridge. The company focus is always to achieve original design, tailored to the client's requirements. Based in South Wales, D.A.R. Design was established by Andrew Rowe in 1990 following his training in three-dimensional design in metal. The Delta Street commission was a collaboration with artist Simon Fenoulhet, who specialises in making light sculptures.

ipv Delft

ipv Delft is a Dutch engineering, architecture and design office that specialises in street furniture, lighting and bridges. The company's vision is that a good design should be simple, effective and inspired. That way, products will fit the urban fabric today and in the future.

Knowledge, realism and imagination are at the core of all of ipv Delft's activities. With sensible choices and smart solutions, the office can make successful designs no matter the budget. ipv Delft was established in 1996 and has since grown into a company with a team of twenty designers, architects and engineers.

Stephen Diamond Associates

STEPHEN**DIAMOND**ASSOCIATES
CHARTERED LANDSCAPE ARCHITECTS

Stephen Diamond Associates is a progressive design-orientated landscape architecture consultancy based in Dublin. Since its inception in the summer of 2000 the practice has been involved in a diverse body of inventive work at various scales throughout Ireland.

Their approach to landscape architecture is site-generated, with careful consideration given to the history, geology, ecology, microclimate and landscape of each site and its context. They place emphasis on creativity and rigorous conceptual development in their search for robust and timeless design proposals.

Gustafson Guthrie Nichol

Gustafson Guthrie Nichol (GGN) is a landscape architecture practice based in Seattle, Washington. Founded by partners Jennifer Guthrie, Kathryn Gustafson, and Shannon Nichol, GGN works throughout the Americas and Asia. Its projects are developed by designers with professional backgrounds in landscape, architecture, engineering, graphics, ecology, and other sciences, and express the sculptural qualities of contextual landscape. GGN offers experience in designing high-use landscapes in complex, urban contexts. GGN's landscapes are widely varied in type and scale, but they share qualities as healthy settings for diverse and ever-changing activities. The landform of each space is carefully shaped to feel serenely grounded in its context and comfortable at all times – whether bustling with crowds, offering moments of contemplation, or doing both at once.

Kathryn Gustafson is also a partner in the UK design firm Gustafson Porter. In 2012, Gustafson was awarded the Arnold W. Brunner Memorial Prize in Architecture from the American Academy of Arts and Letters.

Barbara Grygutis

Barbara Grygutis is an award-winning, U.S.-based sculptor and public artist whose work is in permanent public art collections throughout the United States and beyond, including Miami, Florida; Philadelphia, Pennsylvania; New York City; Washington DC; Denver, Colorado; Seattle, Washington; Columbus, Ohio; St. Paul, Minnesota; Calgary, Alberta, Canada; Phoenix, Arizona, and in Tucson, Arizona, where she lives and works. Light, natural daylight and artificial light, is integral to much of Barbara Grygutis' work. With her formal training in architecture and fine art and an on-going love of historic preservation, public art combines her passions. The artist's interest in nature is evident, and though the natural world provides the genesis of her working vocabulary, such forms are significantly altered through concept development, scale, and materials. Her works of art create a place of reflection, where the beauty of the natural world can be seen in the built environment.

Diana Cabeza

Diana Cabeza is an architect who studied at University of Belgrano, Buenos Aires, and graduated with honours from the Prilidiano Pueyrredon National School of Fine Arts.

Cabeza is a designer specialising in urban equipment, driven by re-thinking and designing urban elements and supports for community use in the public space. She is the principal designer and president of Estudio Cabeza Urban Elements for Public Spaces.

Her designs have been published in the following magazines: Domus, Abitare, The Plan, Nexus, Visions, Interior Design, Architectural Record, Summa + and Barzón as well as in books published abroad. Her designs have been licensed out to foreign companies in Europe and Latin America. Her urban pieces furnish public spaces in the cities of Buenos Aires, Córdoba, Puerto Madryn, Mendoza, Tokyo, Zurich, Paris, Washington DC, New York, Miami, Perth to name a few.

She obtained the Konex Platinum Award in industrial design and four Good Design Awards in 2012; the DARA Award in 2010; the first place in the National Competition of Urban Furniture for the city of Buenos Aires GCBA/SCA together with Wolfson/Heine in 2004; the ICFF Editors Award New York 2003 and the Konex Merit Award in 2002.

Pepe Gascón Arquitectura

Established in 2003, Pepe Gascón Arquitectura is a multidisciplinary consulting architecture that has a young and innovative team, which develops projects of architecture, urbanism and interior design. His work is oriented to both the public and private, and includes projects of varying scale and technical complexity.Its work has been published and exhibited in numerous publications and exhibitions, national and international.

Awards:
2012
The International Architecture Award Shortlist
Emirates Glass LEAF Awards
Selected at the awards Premios ARQUIA Próxima (3rd edition, 2010-2011)
Selected at the awards Premios Arquitectura en Positivo (CSCAE)
2010
Silver medal at the International Design Awards IDA'11 2010
Costa Rica International Architecture Biennial
Selected at the 6th Rosa Barba European Landscape Prize
2009
Gold medal at the 2009 Miami + Beach Biennial 2008
Selected for the Premis Catalunya Construcció
2007
Finalist at the 4th Bienal d'Arquitectura del Vallès (4BAV)

Alexander Lotersztain

Alexander Lotersztain was born in Buenos Aires, Argentina in 1977; he graduated from Design at Griffith University QCA in 2000. He is director of Derlot Pty. Ltd., a multidisciplinary studio focusing on projects including product, furniture, branding, hotel design, interior design and art direction with clients both nationally and internationally. Derlot Editions is co-brand of derlot and produces a range of Australian made furniture and lighting for the contract and domestic markets and distributed worldwide.

Had participated in international exhibitions with Sputnik, Designers Block London, Tokyo, Milano, New York, San Francisco, Berlin, Moscow and of his products is part of the design Collection at the Pompidou Museum in Paris.

Mr. Lotersztain won the Inaugural Queensland Premier's Smart State Designer of the Year Fellowship Award 2010, and recently returned to judge the award for 2011. He was named one of 100 most influential top designers worldwide in &fork by Phaidon, top 10 most influential faces in Design by Scene Design Quarterly 2007 and top 10 of 100 Young Brightest Australian Achievers Bayer/Bulletin Award. He has won many awards in both product and interior design and his work has appeared in design journals around the world. Alexander is also part of the "Smart State Design Council" for the Queensland Government in Australia, drafting the Smart State design Strategy for 2020.

David Brooks

David Brooks was born in 1975 in Brazil, Indiana, USA. He studied art at the Städelschule, Staatliche Hochschule für Bildende Künste in Frankfurt am Main, Germany, and The Cooper Union, School of Art, New York and completed his studies at Columbia University, School of the Arts, New York.

Brooks presented his work "Preserved Forest" (2010) in MoMA PS1's 2010 Greater New York show. In 2011, David Brooks showed his critically acclaimed "Desert Rooftops", as part of an Art Production Fund commission in the Last Lot in Times Square. Brooks's forthcoming exhibitions include solo shows at the American Contemporary (New York) and the Galerie für Landschaftskunst (Hamburg). Brooks lives and works in New York City.

Brooks is best known for his work which considers the relationship between the individual, manmade and natural environments. His recent commissions for the Cass Sculpture Foundation, "Picnic Grove" and "Sketch of a Blue Whale (enlarged to scale 23m, 154 tonnes)" combine the natural world through subject matter or materials with everyday infrastructures to call to mind the proximity of such relationships.

ModelArt Studio

The Group is engaged in the design and manufacture of utility installations, participation in competitions, exhibitions and professional workshops, development of physical models and the like.

WOW Architects | Warner Wong Design

WOW is an international consultancy offering professional design services in architecture, interior design, landscape design and master planning.

An award-winning firm recognised for extensive experience in cross-border services, their projects span the Asian region and extend as far as the Middle East and East Africa. Their core competency lies in the areas of Hospitality, Residential and Commercial design with strong lifestyle-driven concepts. Established as Warner Wong Design in the year 2000 by Wong Chiu Man and Maria Warner Wong, WOW has since evolved into a dynamic design group comprising Warner Wong Design, WOW Architects Pte Ltd, WOW DLab, WOW Design and WOW Design Studio.

Dan Corson

Dan Corson is an artist working nationally and internationally and maintains a studio (Corson Studios LLC), based in Seattle, Washington. Corson's artwork straddles the disciplines of Art, Theatrical Design, Architecture, Landscape Architecture and sometimes even Magic. His award-winning projects have ranged from complex rail stations and busy public intersections to quiet interpretive buildings, meditation chambers and galleries. Corson's work is infused with drama, passion, layered meanings and often engages the public as co-creators within his responsive and immersive art environments.

J. MAYER H. und Partner

J. MAYER H und Partner, Architekten focuses on works at the intersection of architecture, communication and new technology. Recent projects include the villa Dupli Casa near Ludwigsburg, Germany and Metropol Parasol, the redevelopment of the Plaza de la Encarnacion in Sevilla, Spain, the residential building JOH3 in Berlin, Germany, and several public and infrastructural projects in Georgia, for example an airport in Mestia, the border checkpoint in Sarpi and two rest stops along the highway in Gori. From urban planning schemes and buildings to installation work and objects with new materials, the relationship between the human body, technology and nature form the background for a new production of space.

J. MAYER H. was founded in 1996 by Jürgen Mayer H. in Berlin. In January 2014, Andre Santer and Hans Schneider joined as partners in the firm. Jürgen Mayer H. studied at Stuttgart University, The Cooper Union and Princeton University. His work has been published and exhibited worldwide and is part of numerous collections including MoMA New York and SF MoMA and also private collections. National and international awards include the Mies van der Rohe Award, Emerging Architect, Special Mention 2003; Winner of Holcim Award Bronze 2005; and Winner of Audi Urban Future Award 2010. Jürgen Mayer H. has taught at Princeton University, University of the Arts Berlin, Harvard University, Kunsthochschule Berlin, the Architectural Association in London, the Columbia University, New York and at the University of Toronto, Canada.

ah asociados

ah asociados is an architectural company established in 1995 with branches in Pamplona, Bilbao, Barcelona, Madrid, Doha and Panama. Its main purpose was to provide technical services for Planning, Design, Site Control and Project Management, after years of experience in housing, public facilities and urban design.

The ah group, based upon a standard organisational framework, is focused on inter-disciplinary team work to obtain better results, adapted to the identity of each client, with the quality of a personalised office and the efficiency of a large company. Now ah asociados has specialised teams in the design and execution of collective housing; industrial facilities; public centres for culture, education and health; architectural renewal; urban design; structural design; technical specifications; site control; quantity surveyors and project management.

Its work has been awarded with prizes in many local and international competitions and has been widely published in architectural exhibitions magazines and lectures including University Forums, Schools of Architecture and Professional Institutes.

Allegory Studio

Allegory Studio is Swiss architecture and design firm based in Geneva. It was founded in 2010 by Albert Schrurs (EPFL grad. Architect + ECAL MAS Luxury grad. Designer), after having worked for Louis Vuitton as an Architect and for several important firms such as Shigeru Ban, Kengo Kuma and Ma Yansong, in Paris, Tokyo and Beijing.

Miloš Milivojević

Miloš Milivojević was born in Belgrade, in April 1985. He obtained his Master diploma as the best student in his generation in 2009 from the Faculty of Architecture at the University of Belgrade. Curently, he works as a freelance architect and as a teaching associate at the Department of Architecture, Faculty of Architecture, University of Belgrade. He obtained awards and prizes for his works, which were presented at various national and international exhibitions. Miloš and photographer Marko Sovilj founded the fashion brand MIMIMASO in 2010.

Ian McChesney

Ian McChesney works as an independent architect, designer and sculptor. Commissions include "Out of the Strong Came Forth Sweetness" in London's Angel Building, rotating wind shelters for Blackpool's promenade, an award-winning park pavilion in Preston and a range of small batch-produced lamps. He has recently completed Arrival and Departure for the grounds of Plymouth University in Devon.

+uniformarchitects

+ uniformarchitects

Young dynamic team with their own design facilities supported by reliable subcontractors. They have gained experience in renowned Slovak and foreign studios and years of their own practice. +uniformarchitects focus on quality creative design of modern residential, civil and industrial architecture, including interior and industrial design as well as reconstruction and rehabilitation of facilities and historical monuments.

Mathieu Lehanneur

Paris-based designer Mathieu Lehanneur has a vast repertoire. His creations range from sensory designs that interact with other objects, to futuristic environments that soothe the human psyche. Lehanneur was one of the first designers to successfully combine organic materials with digital technology, resulting in designs that use live organisms to purify the air or imbue built structures with living parts. Graduated from ENSCI-Les Ateliers in 2001, Lehanneur opened that same year his own studio dedicated to industrial design and interior architecture.

In 2006, he got the Carte Blanche from the VIA and he was awarded the Grand Prix de la Création from the city of Paris. In 2008, he received the Talent du Luxe Award and the Best Invention Award (USA) for Andrea, its air-cleaning system using plants.

His studio worked for many prestigious clients such as: Nike, Issey Miyake, Cartier, Biotherm, Veuve Cliquot, Poltrona Frau, Yohji Yamamoto, Centre Pompidou, Harvard University, Nespresso, and Schneider Electric.

Oglo

Oglo was founded in september 2009 by Arnaud Dambrine and Emmanuel de France. The sensitive research of shapes, the poetic and symbolic aspect of space and the materiality are in the core of their creative process. Today, they are developing a reflexion about Architecture and Landscape, from small and private scale to urban and public scale.

Arnaud Dambrine, born in 1984 in Paris, is graduated from the Ecole Nationale Supérieure d'Architecture de Versailles. He also has studied in Rome in the Università degli Studi La Sapienza from 2006 to 2007. He worked in Amsterdam for Snitker/Borst/Architecten and in Paris for the office Plan 01 and for ANMA, Agence Nicolas Michelin et Associés.

Born in 1983 in Nouméa, Emmanuel de France is graduated from the Ecole Nationale Supérieure d'Architecture de Versailles. In 2007, he has studied in the Facultad de Arquitectura, Diseño y Urbanismo of Buenos Aires. Afterwards he has worked for the porteño offices Dujovne-Hirsch and Ramos. He has completed his studies with a final project about the interrelation between the Argentina's capital and the River Plate. In Paris, he has worked in free-lance for Emmanuel Vialar Architect.

Jair Straschnow and Gitte Nygaard

Off-ground is the brainchild of award-winning designers Jair Straschnow and Gitte Nygaard. Jair Strashnow and Gitte Nygaard have been living and working in Amsterdam, The Netherlands since 1999. They both run independent studios and have been working together on various projects in public space all over the world – Johannesburg, Durban, Rio de Janeiro, Brussels and Amsterdam to name a few.

In 2009 they were nominated for the Rotterdam Designprijs for their project for street children in Durban, South Africa. Jair was the winner of the furniture category of the Brit Insurance Designs of the year awards at the London Design Museum 2010, and Gitte was the winner of Design and Craft Biennale 2013 in Copenhagen.

SUBARQUITECTURA (ABR 05)

SUBARQUITECTURA (ABR 05) is composed by Andrés Silanes (NOV 1977), Fernando Valderrama (AGO 1979) & Carlos Ba.ón (OCT 1978). They are Architects by the University of Alicante (OCT 04) and Master of Complex Architectures by the University of Alicante (2005 -2007).

They have built the Tram Stop and Plaza on the trafic roundabout of Sergio Cardell (2006) , which was awarded at the IX Spanish Bienal of Architecture, Nominated to the Mies van der Rohe Awards 2009, Finalist at the Valencian Community Awards 2008, Second Prize at Lamp Lighting Sollutions 2008, Honourable Mention at the Balthasar Neumann Prize in Germany, and Selected for the 10 Best Designs in the World by the London Design Museum (2008). Also they have constructed various sports facilities, like the Sports Pavilion in Pedreguer, the Hercules C.F. remodelling and the 3D Athletics Track in Alicante (2010), which has been recently awarded with the Silver Medal and the Accesibility Distinction by the Internacional Olympic and Paralympic Committees. On 2010, Subarquitectura Architects were nominated by Yosihiaru Tsukamoto (from Atelier Bow-Wow) for the Iakov Chernikhov International Prize.

FTL Design Engineering Studio

FTL is a leading architectural design and engineering firm, offering a range of design and consulting services for innovative buildings.

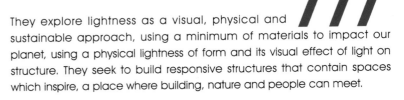

They explore lightness as a visual, physical and sustainable approach, using a minimum of materials to impact our planet, using a physical lightness of form and its visual effect of light on structure. They seek to build responsive structures that contain spaces which inspire, a place where building, nature and people can meet.

后 记

　　本书的编写离不开各位设计师和摄影师的帮助，正是有了他们专业而负责的工作态度，才有了本书的顺利出版。参与本书的编写人员有：

Mark A. Reigelman II, Tonkin Liu, Manuel Ruisánchez, Escofet, Anna Pericas, Shlomi Almagor, FÜNDC, César Gª Guerra © FÜNDC, Ignacio Ciocchini,Marco Castro, Forms+Surfaces, STORE MUU design studio, Daisuke Ito, mmcité, Keha3, Springtime, Jair Straschnow ,Gitte Nygaard, LODEWIJK BALJON landscape architects, Lodewijk Baljon, Smits Rinsma, Bjarke Ingels Group, Studio MA.A&D, Daily tous les jours, Olivier Blouin, Moradavaga, Manfred Eccli, Pedro Cavaco Leitão, TORAFU ARCHITECTS, Fuminari Yoshitsugu, AND-RÉ, ARCHITECTURE RICARDO GUEDES, HUGO PAIVA, NIPpaysage Landscape Architects, Frédérique Ménard-Aubin, Vincent Lemay, Mikyoung Kim Design, Lisa Garrity, Charles Mayer, Karres en Brands Landscape architecture + urban planning, Studio Weave, ModelArt Studio, Alexandre Moronnoz, Le Plan B - Architecture, Lumière, Intérieur, Design, Justin Westover, Nichola Trudgen, D.A.R. Design, Simon Fenoulhet, Barbara Grygutis, Peter Peirce, Stephen Diamond Associates, Gustafson Guthrie Nichol (GGN), ipv Delft, Henk Snaterse, Diana Cabeza, Pepe Gascón Arquitectura, Eugeni Pons, Studio Derlot, Florian Groehn, Igor Solovyov, David Brooks, Barney Hindle, Courtesy of Cass Sculpture Foundation, J. MAYER H. Architects, Nikkol Rot, WOW Architects | Warner Wong Design, Aaron Pocock, C3 Momentum Studio, Corson Studios LLC, Dan Corson, Peter Cook, Uniform Architects, B. Bod'a, M. Fabian, SUBARQUITECTURA, ah asociados, Mathieu Lehanneur, JCDecaux, Oglo, Emmanuel de France, Arnaud Dambrine, Annik Wetter, Allegory Studio, FTL DESIGN ENGINEERING STUDIO, Miloš Milivojević, M.Arch.

ACKNOWLEDGEMENTS

We would like to thank everyone involved in the production of this book, especially all the artists, designers, architects and photographers for their kind permission to publish their works. We are also very grateful to many other people whose names do not appear on the credits but who provided assistance and support. We highly appreciate the contribution of images, ideas, and concepts and thank them for allowing their creativity to be shared with readers around the world.